NATO ASI Series

Advanced Science Institutes Series

A series presenting the results of activities sponsored by the NATO Science Committee, which aims at the dissemination of advanced scientific and technological knowledge, with a view to strengthening links between scientific communities.

The Series is published by an international board of publishers in conjunction with the NATO Scientific Affairs Division.

A Life Sciences B Physics	Plenum Publishing Corporation London and New York
C Mathematical and Physical Sciences D Behavioural and Social Sciences E Applied Sciences	Kluwer Academic Publishers Dordrecht, Boston and London
F Computer and Systems Sciences G Ecological Sciences H Cell Biology I Global Environmental Change	Springer-Verlag Berlin Heidelberg New York Barcelona Budapest Hong Kong London Milan Paris Santa Clara Singapore Tokyo

Partnership Sub-Series

1. Disarmament Technologies	Kluwer Academic Publishers
2. Environment	Springer-Verlag / Kluwer Academic Publishers
3. High Technology	Kluwer Academic Publishers
4. Science and Technology Policy	Kluwer Academic Publishers
5. Computer Networking	Kluwer Academic Publishers

The Partnership Sub-Series incorporates activities undertaken in collaboration with NATO's Cooperation Partners, the countries of the CIS and Central and Eastern Europe, in Priority Areas of concern to those countries.

NATO-PCO Database

The electronic index to the NATO ASI Series provides full bibliographical references (with keywords and/or abstracts) to about 50 000 contributions from international scientists published in all sections of the NATO ASI Series. Access to the NATO-PCO Database is possible via the CD-ROM "NATO Science & Technology Disk" with user-friendly retrieval software in English, French and German (© WTV GmbH and DATAWARE Technologies Inc. 1992).

The CD-ROM can be ordered through any member of the Board of Publishers or through NATO-PCO, B-3090 Overijse, Belgium.

Series F: Computer and Systems Sciences, Vol. 167

Springer

Berlin
Heidelberg
New York
Barcelona
Budapest
Hong Kong
London
Milan
Paris
Singapore
Tokyo

Neural Circuits and Networks

Edited by

Vincent Torre

John Nicholls

International School for Advanced Studies ISAS
Via Beirut n. 2-4
I-34013 Trieste, Italy

Springer

Published in cooperation with NATO Scientific Affairs Division

Proceedings of the NATO Advanced Study Institute
on Neuronal Circuits and Networks, held at the
Ettore Majorana Center, Erice, Italy, June 15-27, 1997

Library of Congress Cataloging-in-Publication Data

Neural circuits and networks / edited by Vincent Torre, John Nicholls.
p. cm. -- (NATO ASI series. Series F, Computer and systems sciences ; vol. 167)
"Proceedings of the NATO Advanced Study Institute on Neuronal Circuits and Networks, held at the Ettore Majorana Center, Erice, Sicily, June 15-27, 1997"--T.p. verso.
Includes bibliographical references.
ISBN 3-540-64929-8 (hc. : alk. paper)
1. Neural networks (Neurobiology)--Congresses. 2. Neural circuitry--Congresses. I. Torre, Vincent, 1950- . II. Nicholls, John, 1929- . III. NATO Advanced Study Institute on Neuronal Circuits and Networks (1997 : Erice, Italy) IV. Series: NATO ASI series. Series F, Computer and systems sciences ; no. 167.
QP363.3.N39 1998
573.8'54--dc21 98-41287
CIP

ACM Subject Classification (1998): F.1.1, C.1.3, I.2.6, J.3

ISBN 3-540-64929-8 Springer-Verlag Berlin Heidelberg New York

Printed in Germany

Typesetting: Camera-ready by editors
Printed on acid-free paper
SPIN: 10691510 45/3142 - 5 4 3 2 1 0

Preface

The understanding of parallel processing and of the mechanisms underlying neural networks in the brain is certainly one of the most challenging problems of contemporary science. During the last decades significant progress has been made by the combination of different techniques, which have elucidated properties at a cellular and molecular level. However, in order to make significant progress in this field, it is necessary to gather more direct experimental data on the parallel processing occurring in the nervous system. Indeed the nervous system overcomes the limitations of its elementary components by employing a massive degree of parallelism, through the extremely rich set of synaptic interconnections between neurons.

This book gathers a selection of the contributions presented during the NATO ASI School „Neuronal Circuits and Networks" held at the Ettore Majorana Center in Erice, Sicily, from June 15 to 27, 1997.

The purpose of the School was to present an overview of recent results on single cell properties, the dynamics of neuronal networks and modelling of the nervous system. The School and the present book propose an interdisciplinary approach of experimental and theoretical aspects of brain functions combining different techniques and methodologies.

Basic properties of single neurons, such as ionic channels, synaptic release and sensory transduction are discussed in the first part; the dynamics of neuronal networks and circuits are analysed in different preparations and with complementary approaches in the second part; models and theoretical approaches for the understanding of how the brain is organized are presented in the third part.

The book is of interest to neuroscientists, molecular biologists, physiologists, psychologists and brain modellers and it is a good example of the interdisciplinary approach required to understand the brain.

Vincent Torre and John Nicholls

Contents

Part 3 Models and Theoretical Approaches

Part 1

Single Neuron Properties

Modal gating of sodium channels and possible physiological implications on hereditary myopathies

Oscar Moran, Mario Nizzari and Franco Conti*

Istituto di Cibernetica e Biofisica, CNR
Via De Marini, 6
I-16149 Genova, Italy

1. INTRODUCTION

Three human genetic syndromes, hyperkalemic periodic paralysis, paramyotonia congenita and potassium aggravated myotonia, have been associated in recent years with mutations of the gene SCN4A, located in chromosome 17q (23.1 to 25.2), which encodes the α-subunit of the voltage-gated sodium channel of the adult skeletal muscle (Barchi, 1995; Cannon, 1996; George, 1995; Hoffman et al., 1995; Lehmann-Horn and Rüdel, 1996). The 17 single-point mutations identified to date in SCN4A are spread over both cytoplasmic and extracellular regions of the polypeptide, some near the voltage sensors, others near the putative inactivation domain, and do not show any obvious clustering pattern associated with distinctive clinical features. A leading pattern is missing also in the correlation of clinical syndromes and electrophysiological properties of the10 mutations studied so far *in situ* (Cannon et al., 1991; Lerche et al., 1993) or in heterologous expression (Cannon and Strittmatter, 1993; Hoffman et al., 1995; Ji et al., 1996; Mitrovic et al., 1995; Mitrovic et al., 1994; Mitrovic et al., 1993). The most obvious abnormalities (notably a slightly slower and incomplete inactivation) can account for muscle hyperexcitability and repetitive firing instead of normal

* Address correspondence to:
Franco Conti, Istituto di Cibernetica e Biofisica, CNR. Via De Marini, 6.
I-16149 Genova, Italy. Email:conti@barolo.icb.ge.cnr.it.

action potentials (Cannon et al., 1993b), and are consistent with the relatively mild and episodic character of the disorders.

Sodium channels have a modal gating behaviour that becomes most apparent when only the major (α) subunit is expressed in frog oocytes (Chahine et al., 1994; Ji et al., 1994; Moorman et al., 1990; Zhou et al., 1991): a slower mode (M2) is seen in this preparation together with the fast mode (M1) that dominates the performance of normal channels *in situ* (Mitrovic et al., 1993; Nilius, 1988) or when α is co–expressed with the auxiliary β_1–subunit (Bennet Jr et al., 1993; Cannon et al., 1993a; Chahine et al., 1994; Isom et al., 1995; Ji et al., 1994; Patton et al., 1994; Zhou et al., 1991). No similar effects on the functioning of the mature channels are known for the β_2-subunit, which seems to favour the expression and insertion of the channels in the membrane (Isom et al., 1995b). Quite apart from the decrease apparently induced by association with the β_1–subunit, an intrinsic increase of the propensity for M2 in mutant α–subunits might be a major cause of hereditary myopathies. A related possibility is that the mutations change the sensitivity of the α–subunit to cellular factors that modulate its modal propensity. The M2-rich expression of α-subunits in oocytes is best suited for verifying these hypothesis. Indeed, we have shown (Moran et al., 1994) that a single mutation of the rat brain sodium channel α-subunit can influence drastically the modal behaviour, quite apart from β1-interactions. Preliminary to this analysis we have undertaken a more detailed characterisation of the relative contribution of the two modes to macroscopic currents mediated by the rat skeletal muscle sodium channel (rSkM1), which is 92% homologous to the human channel (hSkM1), expressed in oocytes by wild type cRNA. We have tested a protocol for dissecting the M1 and M2 components of macroscopic currents, yielding good estimates of their relative probability, and characterised some properties of the channel functioning in each gating mode.

2. MATERIALS AND METHODS

2.1 Sodium channel cRNA expression

The full-length cDNA of the voltage-dependent sodium channel from rat adult skeletal muscle, rSkM1 (Trimmer et al., 1989), was cloned into a high expression recombinant plasmid pGEMHE (Liman et al., 1992). cRNA was generated from the *NheI* linearised plasmid using T7 RNA polymerase (mMessage, Ambiom). After removing template DNA with DNase I, cRNA was extracted, precipitated and stored at –70°C for injection in oocytes.

Xenopus laevis oocytes were isolated enzymalically by treating the frog ovarian tissue with 1 mg/ml of collagenase-A (Sigma) solution. Stage IV and V oocytes were injected with 50 nl of 125-250 ng/μl cRNA DEPC-water solution and incubated 3 to 6 days in Barth solution at 18°C (Stühmer, 1992).

2.2 Electrophysiological recording

Sodium currents were measured from membrane macro–patches in the cell–attached configuration (Hamill et al., 1981; Stühmer, 1992) using an Axopatch-200B patch–clamp amplifier (Axon Instruments). Aluminium–silicate glass micropipettes (Hilgemberg) were coated with silicone rubber and fire polished to a tip diameter yielding a resistance of 0.6 to 1.2 MΩ when the pipette was filled with normal frog Ringer (NFR; in mM: NaCl 115, KCl 2.5, $CaCl_2$ 1.8, HEPES 10; pH = 7.4). The bathing solution in which the oocytes were maintained had the following composition (in mM): KCl 120, TRIS–Cl 20, EGTA 5; pH 7.4. Because of the high potassium concentration of the bath solution, the cell membrane potential was 0 ± 2 mV. Therefore, the membrane patch potential, V, was estimated to be just opposite to the pipette potential. The output of the patch–clamp amplifier was filtered with the low–pass 4–pole Bessel filter built into the patch-amplifier, set at a cut–off frequency of 2 to 10 kHz. The membrane current was sampled at 10 to 50 kHz.

Two–electrodes voltage–clamp technique was used to record whole-oocyte sodium currents (Stühmer, 1992), using a high-voltage amplifier constructed in the laboratory. Recording electrodes of 0.3 to 0.8 mΩ were fabricated from borosilicate glass capillaries (Hilgemberg) and filled with 2M KCl. Oocytes were placed in a temperature-controlled recording chamber filled with normal frog Ringer. The output of the amplifier was filtered with a low–pass four–poles Bessel filter (4304, Ithaco) with a cut–off frequency of 5 kHz. Data were sampled at 20 kHz.

Voltage stimulation and data acquisition were performed by 16 bit D–A and A–D converters (ITC–16, Intrutech), controlled by Macintosh microcomputers, using the Pulse-PulseFit software (Heka Electronik). Linear current responses were measured from sub-threshold stimulations and digitally subtracted. All measurements were done at a controlled temperature of 16 ± 0.5 °C.

3. RESULTS

3.1 Properties of the two modes of gating

The inactivation of the sodium currents recorded in oocytes injected with rSkM1-cRNA have two clearly distinguishable time constants (fig. 1A). The decaying phase of the current can be well fitted with a double-exponential function, yielding a fast time constant τ_1 related with the fast gating mode (M1), and a longer time constant τ_2, corresponding the slow gating mode (M2). The time constants are function of the test potential (fig. 1B), and the ratio τ_1/τ_2 increases from ~17 at -40 mV (τ_1=3.4±2.7, n=8; τ_2=59.4±20.9 ms, n=6; mean±s.d.) to ~52 at -20 mV (τ_1=0.6±0.2 ms, n=8; τ_2=30.3±11 ms, n=6).

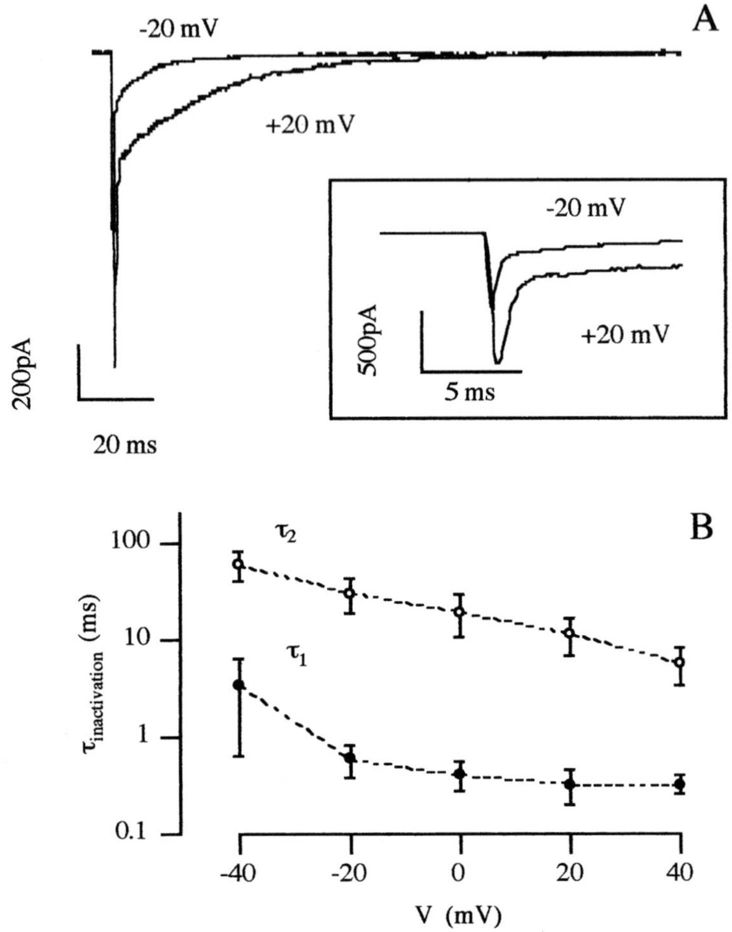

Figure 1. The inactivation of currents mediated by the α-subunit of rat skeletal muscle sodium channels expressed in oocytes is markedly biphasic **A:** Currents elicited by 150 ms test pulses form a holding potential of –120 mV to –20 or +20 mV. A time scale expansion is shown in the inset of the figure. The inactivation phase of the currents can be well fitted by a double-exponential function, yielding a fast component (M1) and a slow component (M2). **B:** The two time constants, τ_1 and τ_2, obtained from similar records at different test potentials, from –40 to +40 mV, are plotted against the potential. Points represent mean±s.d of 4 to 8 experiments.

The recovery from inactivation has different time constants depending on the gating mode. In the experiment shown in fig. 4 on a patch containing channels gating in both modes, a depolarising pulse from the holding potential

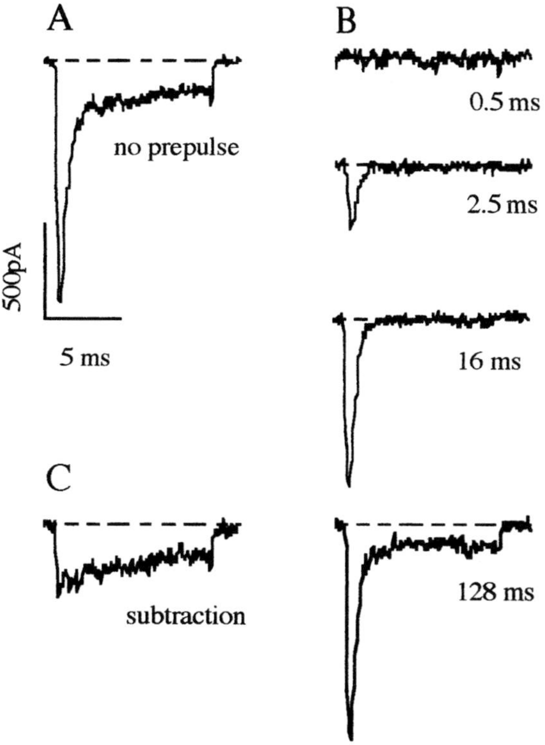

Figure 2. Gating modes have very different rates of recovery from inactivation. **A:** Sodium current with a holding potential of –120 mV and a test pulse to –20 mV showing the two gating modes. **B**: When the test pulse is preceded by a conditioning pulse to 0 mV for 50 ms, followed by a repolarisation to -120 mV of various durations, recovery of M1 is almost complete in ~20 ms, whereas M2 has 50-fold slower recovery. A repolarisation of 0.5 ms is not enough to remove any inactivation; a partial recovery of M1 is obtained for 2.5 ms repolarisation; the recovery of M1 is almost complete in 16 ms, when M2 is still practically undetectable; the inactivation of channels in M2 is abolished only by much longer repolarisations (128 ms). **C:** M2 contribution is obtained by subtraction of the trace in **A** and the 16 ms repolarisation in trace in **B**, that contains only the M1 contribution.

of -120 mV to a test potential of -20 mV elicited a mode M1 current, characterised by the peak being followed by a fast inactivation phase, and an M2 mode current with a slow decaying time constant (fig. 2A). When a 50 ms pre-pulse to 0 mV, followed by a short (0.5 ms) repolarisation at -120 mV preceded the test pulse, most channels, either in M1 or M2, were inactivated during the pre-pulse, and no current was elicited by the test pulse (first trace of fig. 2B).

When the repolarising time was increased up to 40 ms the contribution of M1 increased, as channels gating in this mode recovered from inactivation, while channels in M2 were still inactivated (2nd trace 3rd traces of fig. 2B). The inactivation of M2 channels is removed only by much longer repolarising pulses (last trace on fig. 2B). Subtracting from the unconditioning response the M1 component recorded after 40 ms repolarisation, yield the M2 component (fig. 2C).

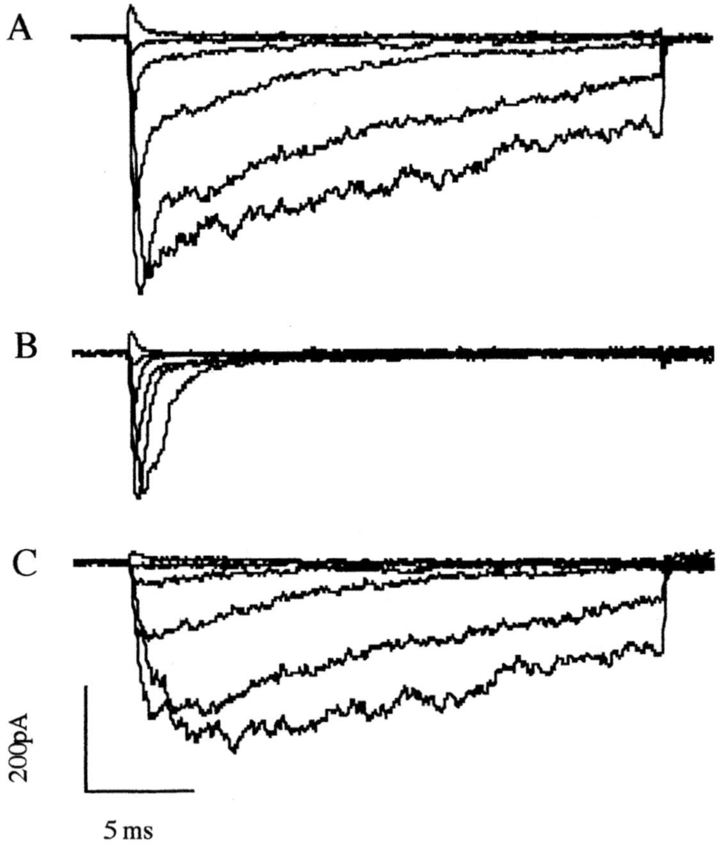

Figure 3. Sodium currents elicited by depolarising voltage pulses contain contributions from channels gating in either one of the two modes (**A**). When the stimulation is preceded by a long depolarising pulse (V=0 mV; t=100 ms), only the currents through channels gating in M1 are recorded after a resting period of 50 ms (V=−120 mV), during which there is almost no recovery from inactivation of M2 (**B**). The subtraction of records obtained without (M1 + M2) and with (M1) conditioning pre-pulses allows the dissection of the gating mode 2 (**C**).

The very different time course of recovery from inactivation of the two modes was exploited to dissect the M1 and M2 components of the current for any voltage stimulation (fig. 3). Indeed, taking advantage of the faster recovery from inactivation of M1 (with a time constant <9 ms), currents with only the M1 contribution can be recorded using a conditioning depolarising pulse followed by a relatively short (~40 ms) repolarising period (fig 3B), during which M2 gating channels do not recover from inactivation (time constant >160 ms). Successively, the M2 contribution to the sodium current can be obtained by subtraction of the current measured with the conditioning pulse protocol from the current recorded without conditioning pre-pulse (fig. 3C). We have applied this simple strategy to dissect current records with a single-gating mode and characterise the voltage–dependent properties of the two modes.

The voltage dependence of the steady state inactivation was measured with the classical double pulse protocol (prepulse of variable amplitude followed by a fixed test pulse), and the peak current response to the test pulse was fitted by a simple Boltzmann function:

$$\frac{I_p}{I_{max}} = \frac{1}{1+\exp\left(\frac{(V - V_{hi})}{v_{si}}\right)}$$

where I_p is the peak current, I_{max} is the maximal peak current, V_{hi} is the half inactivation potential and v_{si} is the voltage characterising the slope of the inactivation curve at V_{hi}. In 10 experiments in which M1 was largely predominant from the beginning of the experiment, or as a result of patch "ageing" (see below), the parameters that characterised this mode were not statistically different from those obtained from 6 patches where M2 was present, and was suppressed by a conditioning depolarising pre-pulse. Similarly, the same voltage-dependence of steady-state inactivation was measured in two experiments where only M2 currents were constitutively present, or in 6 experiments in which M2 was dissected by subtraction of "conditioned" M1, demonstrating that the dissection protocol is quite reliable. These experiments yielded, for M1 gating channels V_{hi}=-106.8±10.3 mV (n=16), and v_{si}=7.9±1.7 mV (n=16). Whereas, the M2 channels inactivate at much more positive potentials (V_{hi}=-72.6±13.2 mV; n=8) and a slightly less steep voltage dependence (v_{si}=9.7±1.9 mV; n=8; fig. 4).

The activation properties of the separate modes were calculated from the corresponding peak currents, I_p, elicited at different potentials. Peak activation probabilities, P_a, were estimated from I_p taking into account the intrinsic non linearity of the open channel currents due to the rectification arising from the asymmetric sodium concentration and due to the voltage-dependent block of the channels by external Ca^{2+} (Armstrong and Cota, 1989; Pusch, 1990). The latter effect was described according to our own measurements (Boccaccio et al.,1998)

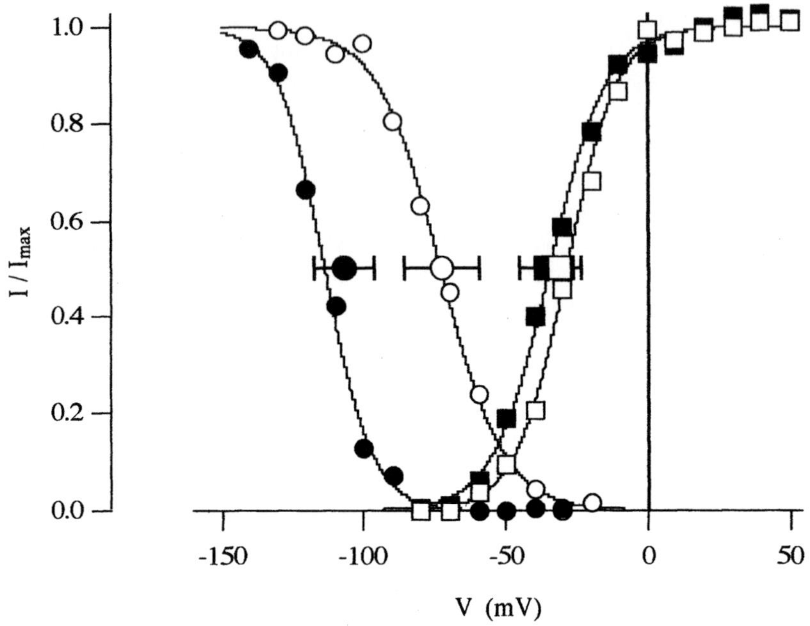

Figure 4. Voltage-dependent properties of activation (squares) and inactivation (circles) of sodium currents gating in M1 (filled symbols) or M2 (empty symbols). Continuous lines are the least-squares fit of experimental data with the equations described in the text. The bigger symbol on each curve represents mean values of V_{ha} and V_{hi} obtained from different experiments, and the horizontal error bars are the corresponding s.d. Notice that the mean values of V_{ha} are not significantly different, while the difference of the mean values of V_{hi} is statistically significant, being M2 about 30 mV more positive.

using an apparent Ca^{2+} dissociation constant at 0 mV of 17 mM and a voltage dependence with a slope-voltage of 42 mV. The Na^+ gradient rectification was described using the Goldman-Hodgkin-Katz equation (Hille, 1992). The voltage-dependence of activation was then evaluated according to a simple Boltzmann function:

$$P_a = \frac{1}{1 + \exp\left(\dfrac{-(V - V_{ha})}{v_{sa}}\right)}$$

where V_{ha} is the half activation potential and v_{sa} is the e-fold voltage-dependence. Similar activation properties of the two modes were estimated from experiments where pure M1 (n=10) or pure M2 (n=8) were recorded or from experiments where the two gating modes were dissected according to the protocol illustrated in fig. 3. In general V_{ha} was 2 to 5 mV more positive for M2 than M1 when activation curves obtained for M1 and M2 were obtained from the same set of records by dissection of the two modes. However, the mean overall values were not significantly different, yielding values of -36.0±9.0 mV (n=17) and -31.6±7.3 mV (n=15) for M1 and M2 respectively. Similarly, v_{sa} was systematically smaller by 2 to 3 mV in the dissected records, but the mean values of 10.6±2.9 mV (n=17) and 6.5±2.0 (n=15) for M1 and M2 respectively, were not significantly different (fig. 4).

3.2 Variability of modal properties

In agreement with previous studies of the modal gating of sodium channel α-subunits expressed in oocytes, we find that the expression of rSkM1 α-subunits produces variable proportions of the two gating modes. Examples of the variability of the modal contribution to the total currents are presented in figs. 5A, 5B and 5C. This variability has been observed in oocytes from different batches. Within 3 minutes from patch formation, in most of the patches (33 out of 67) we observed a mixture of the two modes with prominence of M2 (fig. 5B) or pure M2 (fig. 5C; 26 patches). Records with pure M1 were found in 8 patches (fig 5A). A prominence of M2 was also found in all whole-oocyte currents (23 experiments) recorded after 3 minutes from the beginning of the voltage-clamp of the cell.

The relative probability to be in M2 was evaluated by the ratio of the current after 10 ms, I_{ss}, where most of the current is due to M2, to the peak current, I_p, contributed by both modes. The probability to be in M2 slightly decreases with time after cRNA injection. The parameter I_{ss}/I_p measured in whole-oocytes was 0.70±0.12 (n=10) the 3rd day after cRNA injection, and decreased to 0.6±0.12 (n=7) at the 5th day. The initial records from cell-attached membrane patches showed the same trend with the age of the oocyte. However, while the relative proportion of the two modes did not change for at least 20 minutes when measured in a whole–oocyte, the M1 contribution to the current increased invariably with time at the expense of the M2 component during a patch-recording experiment (fig. 5D). The value of I_{ss}/I_p was 0.59±0.25 (n=30) between the 3rd and the 5th min after the patch formation, and decreased during the first 10 min to 0.43±0.22 (n=20). No further statistically significant increase of the M1 contribution was observed after 10 min of patch formation. This increase in M1 and the decrease in M2 is faster and more prominent when the patch of membrane is excised in the inside-out configuration (fig 5E).

It has been suggested that the gating mode variability is due to intracellular factors that modulate the equilibrium between the gating modes (Nilius, 1988; Zhou et al., 1991). One obvious candidate for such modulation would be a redox reaction occurring in the region of the sodium channel polypeptide involved in the inactivation mechanism, in analogy with observations reported for the

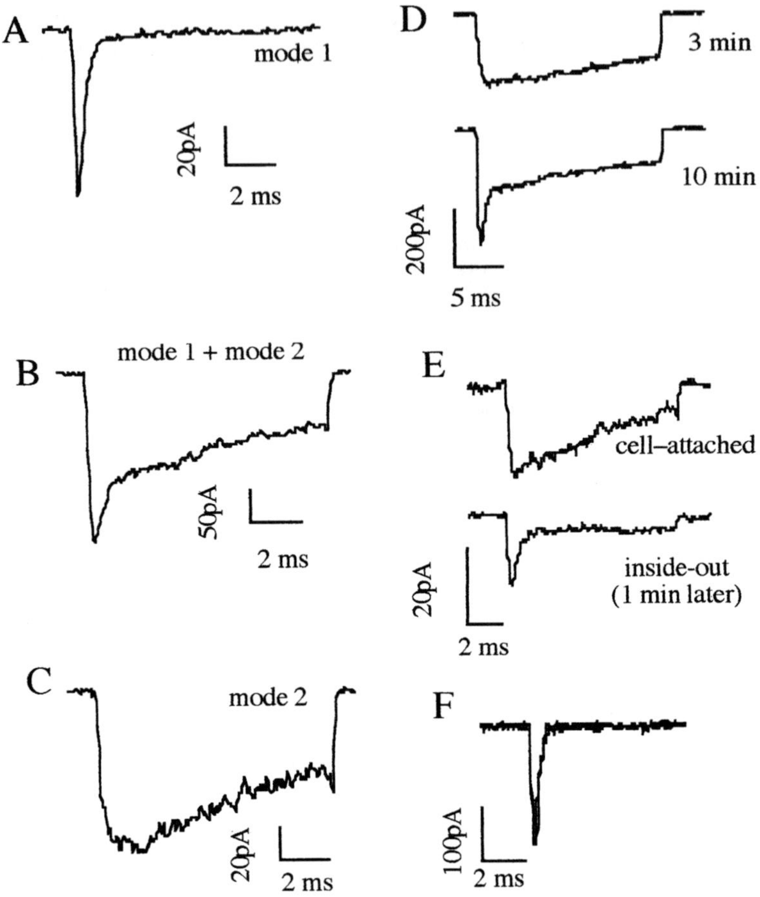

Figure 5. Variability of the relative contribution of the two gating modes in different experiments. **A, B, C:** Sodium currents measured form three different patches, with a holding potential of –120 mV and a test pulse to –20 mV, yielding different contribution of M1 and M2 to the total current. Modal gate change with time during patch clamp experiments. **D**: Sodium current records obtained from the same patch show slow variation of modal gating. In the experiment illustrated here a marked increase of M1 and a concomitant reduction of M2 was clearly observed in about 10 min. **E**: Similar, but much faster, variation of the M1 and M2 contributions is observed in another patch upon its excision in the inside-out configuration. Antioxidant agents do not recover M2. **F**: Records obtained from an inside-out patch, with a holding potential of –120 mV and a test pulse to –20 mV, continuously perfused with the bathing solution, and after 3 min perfusion with bathing solution + 0.5 mM dithiotritol. The two traces are almost indistinguishable.

Shaker potassium channel (Ruppersberg et al., 1991). However, redox reactions do not seem to be involved in the modulation of the modal propensity, because we found no obvious effects from the intracellular or extracellular application of antioxidants. In the experiment shown in fig. 5F the patch was in the inside-out configuration, and most of the sodium current arised from M1 after few minutes from excision. The patch was continuously perfused with the normal bathing solution, and successively with bathing solution + 0.5 mM dithiotritol (DTT). The control current is indistinguishable from the current recorded after 2 min of DTT perfusion.

4. DISCUSSION

We have undertaken a thorough investigation of the modal gating of adult rat skeletal muscle α–subunits expressed in *Xenopus* oocytes. At a first glance, the most obvious distinction between the two gating modes is the difference, of more than one order of magnitude, in the inactivation time constants. Similar observations have been already reported for the same sodium channel isoform (Chahine et al., 1994; Ji et al., 1994; Makita et al., 1996; Trimmer et al., 1989; Zhou et al., 1991), and for the rat brain sodium channels (Auld et al., 1988; Moorman et al., 1990; Moran et al., 1994; Patton et al., 1994), although the two gating modes of the latter isoform are reported to have less prominent differences in their inactivation time constants. Differently, sodium channels cloned from cardiac muscle (Gellens et al., 1992; Makielski et al., 1996; Makita et al., 1996) or from denervated or immature skeletal muscle (Ji et al., 1994) seem to have only the fast gating mode, as their inactivation is usually well fitted by a single exponential. The different inactivation time constants can be correlated with previous single-channel observations, where the unitary events are clustered at the beginning of the test pulse when a channel is gating in the M1 mode, whereas repetitive openings are observed during the test pulse when it is gating in the M2 mode(Moorman et al., 1990; Zhou et al., 1991).

To analyse the voltage-dependent properties of the two gating modes, we have developed a stimulation protocol, that allows the dissection of their contribution to the total currents. This "dissecting" protocol, consisting of a conditioning depolarising pre-pulse followed by a relatively short repolarising period, allows to suppress the M2 contribution from the total current, based on its very slow rate of recovery from inactivation (see fig. 2). The currents obtained in this way have only M1 contribution, and the M2 contribution is obtained by subtracting them from the total current (without conditioning prepulse; see fig. 3). We have validated this protocol by comparing the voltage-dependent properties of these currents with those obtained from patches where only one mode was present. Similar results have been obtained in preliminary experiments studying the modal properties of rat brain sodium currents (Baglietto, 1993).

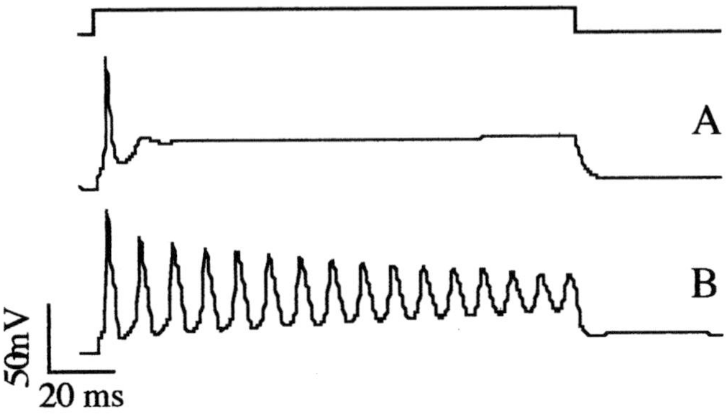

Figure 6. Muscle fibre action potentials were reconstructed by numerical integration of the Hodgkin-Huxley equations, including accumulation of K^+ in the T-tubule space, according to the model proposed by Cannon et al. (1993). All parameters were similar to those used by Cannon and collaborators, except that M2 channels were introduced to the model, with the same voltage dependence of activation of M1 channels, but with a slower (35-fold) and voltage-shifted (+30 mV) inactivation. In control conditions, with all sodium channels gating in M1 (**A**), a long (150 ms) stimulus of 150 μA/cm^2 evokes a single action potential spike. When a small fraction of channel (5%) is gating in M2 (**B**), the same stimulus produces a repetitive firing of action potentials, similar to those occurring in muscle fibres of patients with hereditary single-point sodium channel mutation myopathies.

4.1 Voltage dependence of the gating modes ant its implication in hereditary myopathies

The gating modes of the sodium channels have a similar voltage–dependence of activation (fig. 4). Differently, a significant difference in the voltage-dependence of steady-state inactivation has been found (fig. 4). The V_{hi} of M1 is surprisingly shifted by 34 mV with respect to that of M2. This major difference between the two modes, that has not been stressed in previous literature. If this shift of inactivation occurs also in skeletal muscle sodium channels *in situ*, it would imply that mode M1 is strongly inactivated at the resting potential, increasing the weight of M2 gating in shaping the normal action potentials. The M2 contribution seems to be normally depressed by the modulatory role of the β_1-subunit in natural preparations (Cannon et al., 1993a; Isom et al., 1995; Patton et al., 1994). However, the relative shift of the inactivation curves of M2 and M1 is conserved upon co-expression of the α– and β_1-subunit in frog oocytes (Moran, Nizzari and Conti, unpublished results).

The different V_{hi} of the gating modes may be relevant to explain some of the functional defects that occur in hereditary sodium channel myopathies. Repetitive trains of action potentials and after-stimulus depolarisations, occurring in these diseases and leading to myotonia and paralysis of muscle fibres, have been correlated with the presence of a small fraction of non-inactivating channels. Theoretical reconstructions of the action potentials under these conditions closely resemble the firing behaviour of human muscle fibres in hereditary myotonias associated with point mutations of the skeletal muscle α-subunit sodium channels (Cannon et al., 1993b). As expected, we have verified that the model of Cannon et al. (1993) predicts the same phenomenology also when a small percentage of sodium channels is in the M2 gating mode (fig. 6). Thus, some of the mutations that affect the physiology of the muscle sodium channels could do so by simply altering the intrinsic modal propensity of the α-subunit, and this possibility is the subject of our present investigations.

4.2 Variability of the modal gating may reflect modulatory factors

A puzzling characteristic of the modal gating of sodium channels expressed in oocytes is the large variability in modal propensity. To some extent, this could reflect a variability of the presence or activity of intracellular regulation factors in different oocytes. This hypothesis is supported by the decay of the relative M2 probability, with time after cRNA injection and by the faster reduction of M2 when a patch of membrane is excised (see fig. 5E). Among possible candidates for such regulation, there are redox reactions, phosphorilation of the sodium channel protein or interactions with G-proteins. We have explored possible role of redox reactions in the modulation of the modal gating, similar to that observed for potassium channels (Ruppersberg et al., 1991) and for sodium channels in the Ranvier node of rat (Mitrovic et al., 1993), but we must discard this type of mechanism in our preparation (fig. 5F). Although modulation of brain and cardiac muscle sodium channels by protein kinase-A has been described (Li et al., 1992; Wickman and Clapham, 1995), it is unlikely that this metabolic pathway is involved in the regulation of the gating modes, since its effect is mostly on the magnitude of the current, without any significant effect on the current kinetics or voltage dependence. Furthermore, the skeletal muscle sodium channel lacks four of the six putative phosphorilation sites present in the brain or heart channels, rendering this possibility more remote. More important could be the protein kinase-C pathway, that has been demonstrated to modulate the inactivation kinetics of brain sodium channels (Numann et al., 1991). There is extensive literature about a direct modulation by G-proteins of ionic channels, in particular calcium channels (for review see (McDonald et al., 1994; Wickman and Clapham, 1995), and few reports suggest this mechanism as a possible candidate for sodium channel modulation (Cohen-Armon et al., 1989; Ma et al., 1994).

A peculiar characteristic of the modulation of modal gating of skeletal muscle sodium channels is the decrease of M2 vs. M1 propensity during a patch experiment. Such effect has been also observed in single-channel recordings (Moorman et al., 1990; Zhou et al., 1991). This phenomenon is not found when currents are recorded from the whole-oocyte, strongly suggesting that the patch formation alters the properties of the channels in the membrane region under the recording patch pipette. It is plausible that the mechanical stress of the patch has an indirect effect on some modulatory factors, or a more direct effect on the possible channel-cytoskeleton interactions. The hypothesis that the disruption of the cytoskeleton in contact with the patched membrane may unfavor the M2 gating mode is attractive, but is in apparent disagreement with previous observations of the opposite effect, namely a slowing of the sodium current inactivation time constants consequent to chemical destabilisation of the cytoskeleton (Udovina et al., 1995). This type of experiment is worth of being repeated in our preparation.

ACKNOWLEDGMENTS

We thank E. Gaggero for the construction of the voltage-clamp amplifier, G. Gaggero and D. Magliozzi for the construction of the recording chambers. This work was supported by Telethon-Italy (Grant 926).

REFERENCES

Armstrong, C.M. and Cota, G. 1989. Calcium ion as a cofactor in Na channel gating. *Proc. Natl. Acad. Sci. U.S.A.* 88:6528-6531.

Auld, V.J., Goldin, A.L., Krafte, D.S., Marshall, J., Dunn, J.M., Catteral, W.A., Lester, H.A., Davidson, N. and Dunn, R.J. 1988. A rat brain Na^+ channel α subunit with novel gating properties. *Neuron* 1:449-61.

Baglietto, L. 1993. Studio dei fenomeni di inattivazione del canale al sodio espresso in ovociti di rana. Tesi di Laurea in Fisica. Facoltà di Scienze Matematiche Fisiche e Naturali. Università di Genova, Genova.

Barchi, R.L. 1995. Molecular pathology of the skeletal muscle sodium channel. *Ann Rev Physiol* 57:355-385.

Bennet Jr, P.B., Makita, N. and George Jr, A.L. 1993. A molecular basis for gating mode transitions in human skeletal muscle Na^+ channels. *FEBS Lett.* 326:21-24.

Boccaccio, A., Moran, O. and Conti, F. 1998. Calcim dependent shifts of Na+ channel activation correlated with the state dependence of calcium-binding to the pore. *Eur. Biophys. J.* in press

Cannon, S.C. 1996. Sodium channel defects in myotonia and periodic paralysis. *Ann. Rev. Neurosci.* 19:141-164.

Cannon, S.C., Brown, R.H. and Corey, D.P. 1991. A sodium channel defect in hyperkalemic periodic paralysis: potassium-induced faliure of inactivation. *Neuron* 6:619-626.

Cannon, S.C., Brown, R.H. and Corey, D.P. 1993b. Theoretical reconstruction of myotonia and paralysis caused by incomplete inactivation of sodium channels. *Biophys. J.* 65:270-288.

Cannon, S.C., McClatchey, A.I. and Gusella, J.F. 1993a. Modification of the Na^+ current conducted by the rat skeletal muscle a subunit by coexpression with α human brain β subunit.

Cannon, S.C. and Strittmatter, S.M. 1993. Functional expression of sodium channel mutations identified in families with periodic paralysis. *Neuron* 10:317-26.

Chahine, M., Bennet, P.B., George Jr, A.L. and Horn, R. 1994. Functional expression and properties of human skeletal muscle sodium channel. *Pflügers Arch.* 427:136-142.

Cohen-Armon, M., Sokolovsky, M. and Dacal, N. 1989. Modulation of the voltage depepndet sodium channel by agents affecting G-proteins: a study in *Xenopus* oocytes injected with brain RNA. *Brain Res.* 496:197-203.

Gellens, M.E., George Jr, A.L., Chen, L.Q., Chahine, M., Horn, R., Barchi, R.L. and Kallen, R.G. 1992. Primary structure and functional expression of the human cardiac tetrodotoxin-insensitive voltage-dependent sodium channel. *Proc. Natl. Acad. Sci. U.S.A.* 89: 554-8.

George, A.L. 1995. Molecular genetics of ion channel diseases. *Kidney International* 48:1180-1190.

Hamill, O., Marty, A., Neher, E., Sakmann, B. and Sigworth, F. 1981. Improved patch-clamp techniques for high resolution current recording from cells and cell-free membrane patches. *Pflüg. Arch.* 391:85-100.

Hille, B. 1992. Ionic channels of excitable membranes. Sinauer, Sunderland, Mass:.

Hoffman, E.P., Lehmann-Horn, F. and Rüdel, R. 1995. Overwcited or inactive: Ion channels in muscle disease. *Cell* 80:681-686.

Isom, L.L., Ragsdale, D.S., De Jongh, K.S., Westenbroek, R.E., Reber, B.F., Scheuer, T. and Catterall, W.A. 1995b. Structure and function of the beta 2 subunit of brain sodium channels, a transmembrane glycoprotein with CA motif. *Cell* 83:433–442.

Isom, L.L., Scheuer, T., Brownstein, A.B., Ragsdale, D.S., Murphy, B.J. and Catterall, W.A. 1995. Functional coexpression of the β_1 and type IIA α subunits of socium channels un mammalian cell line. *J. Biol. Chem.* 270:3306-3312.

Ji, S., George, A.L., Horn, R. and Barchi, R.L. 1996. Paramyotonia congenita mutations reveal different roles for segments S3 and S4 of domain D4 in hSkM1 sodium channel gating. *J. Gen. Physiol.* 107:183-194.

Ji, S., Sun, W., George Jr, A.L., Horn, R. and Barchi, R.L. 1994. Voltage-dependent regulation of modal gating in the rat SkM1 sodium channel expressed in *Xenopus* oocytes. *J. Gen . Physiol.* 104:625-643.

Lehmann-Horn, F. and Rüdel, R. 1996. Molecular pathophysiology of voltage-gated ion channels. *Rev. Physiol. Biochem. Pharmacol.* 128:159-268.

Lerche, H., Heine, R., Pika, U., George, A., Mitrovic, N., Browatzki, M., Weiss, T., Rivet-Bastide, M., Franke, C., Lomonaco, M., Ricker, K. and Lehmann-Horn, F. 1993. Human sodium channel myotonia: slowed channel inactivation due to substitution of a glycine within the II-IV linker. *J. Physiol.* 470:13-22.

Li, M., West, J.W., Lai, Y., Scheuer, T. and Catterall, W.A. 1992. Functional modulation of brain sodium channels by cAMP-dependent phosphorilation. *Neuron* 8:1151-1159.

Liman, E.R., Tytgat, J. and Hess, P. 1992. Subunit stoichiometry of a mammalian K^+ channel determined by construction of multimeric cDNAs. *Neuron* 9:861-71.

Ma, J.Y., Li, M. and Catterall, W.A. 1994. Modulation of rat brain Na^+ channels by a G-protein coupled pathway. *Proc. natl. Acad. Sci. USA* 91:12351-12355.

Makielski, J.C., Limberis, J.T., Chang, S.Y., Fan, Z. and Kyle, J.W. 1996. Coexpression of β_1 with cardiac sodium channel α subunits in oocytes decreases lidocaine block. *Mol. Pharmacol.* 49:30-39.

Makita, N., Bennet Jr, P.B. and George Jr, A.L. 1996. Multiple domains contribute to the distinct inactivation properties of human heart and skeletal muscle Na^+ channels. *Circ. Res.* 78:244-252.

McDonald, T.F., Pelzer, S., Trautwein, W. and Pelzer, D.J. 1994. Regulation and modulation of calcium channels in cardiac, skeletal and smooth muscle cells. *Physiol. Rev.* 74:365-507.

Mitrovic, N., George, A.L., Lerche, H., Wager, S., Fhalke, C. and Lehmann-Horn, F. 1995. Different effect on gating of three myotonia-causing mutations in the inactivation gate of the human muscle sodium channel. *J. Physiol.* 487: 107-114.

Mitrovic, N., George, A.L.J., Heine, R., Wagner, S., Pika, S., Hartlaub, U., Zhou, M. and Lerche, H. 1994. K(+) aggravated myotonia:destabilisation of the inactivated state of the human muscle Na+ channel by V1549M mutation. *J. Physiol.* 478:395-402.

Mitrovic, N., Quasthof, f.S. and Grafe, P. 1993. Sodium channel inactivation kinetics of rat sensory and motor nerve fibres and their modulation by glutathione. *Pflügers Arch.* 425:453-461.

Moorman, J.R., Kirsch, G.E., VanDongen, A.M.J., Joho, R.H. and Brown, A.M. 1990. Fast and slow gating of sodium channels encoded by a single mRNA. *Neuron* 4:243-52.

Moran, O., Gheri, A., Zegarra-Moran, O., Imoto, K. and Conti, F. 1994. Proline mutations on the S4 segment of brain sodium channel II. *Biochem. Biophys. Res. Comm.* 202:

Nilius, B. 1988. Modal gating behaviour of cardiac sodium channels in cell-free membrane patches. *Biophys. J.* 53:857-62.

Numann, R., Catterall, W.A. and Scheuer, T. 1991. Functional modulation of brain sodium channels by protein kinase C phosphorilation. *Science* 254:115-118.

Patton, D.E., Isom, L.L., Catterall, W.A. and Goldin, A.L. 1994. The adult rat brain β_1 subunit modifies activation and inacivation gating of multiple sodium channel α subunits. *J. Biol. Chem.* 269:17640-17655.

Pusch, M. 1990. Divalent cations as probes for structure-function relationships of cloned voltage-dependent sodium channels. *Eur. Biophys. J.* 18:327-333.

Ruppersberg, J.P., Stocker, M., Pongs, O., Heinemann, S.H., Frank, R. and Koenen, M. 1991. Regulation of fast inactivation of cloned mammalian Ik(A) channels by cysteine oxidation. *Nature* 352:711-14.

Stühmer, W. 1992. Electrophysiological recording of *Xenopus* oocytes. *In* Methods in Enzymology: Ion Channels, B. Rudy and L.E. Iverson. Academic Press, London. 319-339.

Trimmer, J.S., Cooperman, S.S., Tomiko, S.A., Zhou, J., Crean, S.M., Boyle, M.B., Kallen, R.G., Sheng, Z., Barchi, R.L., Sigworth, F.J., Goodman, R.H., Agnew, W.S. and Mandel, G. 1989. Primary structure and functional expression of a mammalian skeletal muscle sodium channel. *Neuron* 3:33-49.

Udovina, A.I., Shander, G.S. and Makielski, J.C. 1995. Cytoskeleton modulates gating of voltage-dependent sodium channel in heart. *Am. J. Physiol.* 269:H203-H214.

Wickman, K. and Clapham, D.E. 1995. Ion channel regulation by G proteins. *Physiol. Rev* 75:865-885.

Zhou, J., Potts, J.F., TrimmerJ.S., W.S., A. and F.J., S. 1991. Multiple gating modes and the effect of modulating factors on the μ1 sodium channel. *Neuron* 7:775-785.

c-FOS PROTEIN AND NADPH-DIAPHORASE DETECTION IN RAT MIDBRAIN AND SPINAL CORD AFTER CONTUSION INJURY

E. A. Del Bel[1] and H. L. A. Defino[2]
1 Department of Physiology, FORP, 1040-904, Ribeirão Preto, SP, Brazil; 2 Department of Orthopedic Surgery, HC-FMRP, 1040-904, Ribeirão Preto, SP, Brazil.

Abstract: Spinal cord injury appears to involve neuroplastic remodeling or cell death that is the result of intense neuronal cells response to lesion. The damage may occur near the site of lesion, as well as in proximal brain areas. Thus, neurons located in brain centers and functionally connected to the damaged areas may present constitutive changes after spinal cord damage. Experimental studies and clinical observations show that spinal cord trauma may be amplified by secondary neuronal damage. However, little is known about the molecular mechanism that initiates and maintains this neuronal reaction. We have started an investigation of neural changes induced by acute spinal cord injury. We assessed the degree of functional neurologic impairment that follows spinal cord injury by behavioral tests and histological methods. In lesioned areas and throughout the central nervous system expression of the transcription factor protein c-Fos was investigated by immunocytochemistry. The localization of c-Fos may indicate, in addition to neuronal activation, spatial and temporal patterns related to the type of stimulus and to the long-term variation in neuronal physiology. In parallel, to identify spinal neurons that synthesize Nitric Oxide, cells and fibers histochemically stained for NADPH diaphorase (a nitric oxide synthase), were studied. Fos expression was detected in NADPH-diaphorase positive cells at intermedio lateral column, central canal, dorsal horn, including nucleus trigeminalis caudalis, two hours after contusion by weight drop. It was found that NADPH-diaphorase is induced in spinal motoneurons, neurons that are normally NADPH-diaphorase negative; however, no Fos protein expression was detected in these cells. The results suggest that central nervous system areas close or proximal to the lesion site, are activated after contusion by weight drop. In addition they show that Fos protein expression and changes on NADPH-diaphorase activity after spinal cord contusion are not necessarily dependent on each other.

Key words: Motoneurons, NADPH diaphorase, c-Fos immunocytochemistry, central canal, dorsal horn neuron, rat, spinal cord injury, lesion site, transcription factor, immediate early gene, c-Fos, functional assessment

Following trauma, pathological glutamate release from depolarized and damaged neurons may result in persistent stimulation of neural glutamate receptors (43, 60). This excessive receptor stimulation associated with calcium increase (30, 65, 81), and other factors, are responsible for secondary effects, which result in progressive neuronal damage after trauma. This is a slow spreading neuronal degeneration termed excitotoxicity (12,13,47). Excitotoxicity is also associated with a cascade of intraneuronal events that depends on intrinsic properties of the neurons as well extraneuronal components (25, 27, 29). All these factors can contribute to the neural cell response to injury such as degenerative, reparative or plastic processes.

Spinal cord injury appears to involve plastic remodeling or cell death, which is the result of intense neuronal cells response to lesion. The damage may occur near the site of lesion, as well as in brain areas proximal to the primary lesion. Thus, neurons located in brain centers and functionally connected to the damaged areas may be functionally activated function after spinal cord damage (10, 11, 37, 50-53, 62). However, little is known about the molecular mechanism that initiates and maintains these effects.

Excitotoxicity events can be associated with selective changes in gene expression (14, 22). Immediate-early genes (IEG), that encode proteins that bind to DNA at sites know to regulate gene expression, are natural candidates to modulate/mediate long-term changes in the nervous system. IEG are expressed rapidly and transitorily in nervous cells after specific types of stimulation, including trauma (25, 27, 29, 50-53, 67). Several studies suggest that the expression of IEG proteins in nervous cells indicates neuronal activation related to the stimulus applied. Among the best studied IEG are the fos (c-fos, fos b, fra-1, fra-2) and Jun (c-Jun, Jun b, Jun d) families (50-52).

The demonstration of a rapid induction of the IEG c-Fos in a subset of spinal neurons as a result of sensory stimulation (38), was perhaps the first indication that IEG are important for long term changes in the central nervous system (CNS). Their response to damage may induce the synthesis of new proteins that will help the process. In addition, transection of the sciatic nerve of newborn animals induced Fos expression in the neuronal population vulnerable to damage (35, 56, 64). Section or crush of the sciatic nerve in rat results in substantial and long-lasting expression of c-Jun in dorsal root ganglion cells and motor neurons of spinal cord (38, 40). However, this increase is seen maximally after 24 hours and is not detectable 2 hours after injury, time of maximum Fos expression. The return of Jun protein to basal concentrations parallels the period for regeneration of the nerve of cell or even cell death itself (14, 22, 35, 38, 40). The induction of a IEG may indicate, in addition to neuronal activation, spatial and temporal patterns related to the type of stimulus and to the long-term variation in neuronal physiology (22, 50-52, 56, 64).

Recently, Nitric Oxide (NO), a free radical gas, has been established as a messenger molecule, with functions related to neuronal plasticity (16, 19, 31, 39, 48, 49, 76). NO is a unique biological effector molecule (16). Under conditions of normal release, NO is a neuronal messenger molecule. With excessive release, NO may function as cytotoxic molecule mediating macrophage-induced cell death as well as being involved in several neurodegenerative process (28, 39, 46, 49). NO is formed from l-arginine by the enzyme NO synthase (48, 49- NOS). NADPH-diaphorase (NADPH-d) staining in many parts of the CNS, reflects NOS activity (17), and has been used as a marker of neurons producing NO (1, 3, 7, 28, 71, 74, 80). Since the cell cannot sequester and regulate the local concentration of NO, the key to regulating NO activity is to control NO synthesis. NOS/NADPH-d is regulated in various neuronal systems in response to neurochemical damage. NO production may be important in spinal cord trauma. For example, NO is synthesized by spinal neurons (7, 8, 54, 61, 77), peripheral axotomy increases the number of spinal motoneurons that produce NO (32, 72-78, 80), suggesting that NO participates in the various processes that occur after neuronal damage. Thus, this activity may be used to analyze responses to aggressive agents by local medullar nervous cells and also by those located far away from the site of damage.

We have been investigating the expression of c-Fos and c-Jun genes to identify specific brain areas activated after stress or seizures (20, 21, 23, 24, 45, 68,69), since these transcriptionally operating proteins are supposed to play a crucial role in cellular growth differentiation and restoration (22, 50-53, 56). We wanted to know whether changes in Fos protein and NADPH-d activity could be related to the initiation of cell body responses of lesioned neurons following spinal cord traumatic lesion (18, 25, 63). This article contains a summary of previously published data together with preliminary original results.

Impact injury is generally considered to provide the most clinically relevant model (63, 73). The injury was produced in the rat with the weight-drop technique. The experimental model used is a modification of that introduced by Allen in 1911 (4, 5) and perfected by Albin and White (2). Male Wistar rats weighting 250-300 g were used. Briefly, animals had free access to water and food and were kept on a 12 h light-dark cycle (lights on at 7:00 am), under controlled temperature conditions. The site of the selected lesion selected was the vertebral segment located between T10 and L1. The animals were anesthetized with tribromoethanol (dose of 2.5 mg/100 g body weight), placed in a stereotaxic apparatus and submitted to the procedure. Animals in the control group (n = 5) were anesthetized and the skin and subcutaneous tissue were divulsedt along the midline to expose the spine. In the laminectomy group (n = 6) the spine was approached along the posterior surface and total laminectomy of the T10-L1 segment was performed so as to expose the dura-mater, which was left intact. In the contusion group (n = 6), the dura-mater was exposed and a lesion was produced by impact on the spine. To produce this lesion a metal disk was placed on the meninges and a 5 g weight was dropped on it by free fall from a height of 17.5 cm inside a teflon tube. After surgical procedure, the muscular and subcutaneous layers and the skin were sutured and the animals were allowed to rest for 2 hours in individual boxes.

Motor function was evaluated clinically according to the classification proposed by Tarlov and modified by Bohlman (8), which is based on the evaluation of motor function of the limbs. Sensorimotor function was then evaluated for the determination and characterization of the extent of spinal injury using the hot plate and the tail-flick tests.

The hot plate test consists of placing the animal on an aluminum plate heated to 55°C. The animals with no nerve damage lick their hind paw or jump on the plate, using for these activities spinal and supraspinal circuits. During the test the animals were left on the hot plate for a maximum period of 60 seconds and removed from it when they did not show any reaction, a fact considered to be the diagnostic criterion of lesion.

The tail-flick test evaluates the integrity of the spinal reflexes of the animal. Each rat was placed in a ventilated glass tube with the tail laid across a nichrome wire coil maintained at room temperature (23 ± 2 °C). The coil temperature was then raised at the rate of 9 °C/s by the passage of electric current, which was adjusted to ensure a tail withdrawal reflex within 2.5 to 3.5 s. A cut-off time of 6 s was established to minimize the probability of skin damage. The animals with no nervous damage flick their tail within 2-3 seconds. The animals that left their tail on the plate for 6 seconds were considered to have spinal damage.

Statistical analysis of the effects of drugs on the hot plate test and tail-flick test was done using multivariate analysis of variance (MANOVA) with repeated measures to compare the groups over all times. The factors analyzed were treatments, time and treatment x time interaction. In case of significant treatment x time interaction, one-way ANOVA followed by the Duncan test was performed for each time. The level of significance was set at $P<0.05$.

Control animals showed normal sensorimotor function, whereas some animals in group submitted to laminectomy showed evidence of spinal cord injury during the procedure. The group submitted to spinal cord contusion showed partial or total paraplegia in all animals studied. The results of the hot plate test showed that the group of animals with contusion injury presented latencies for response (60+/-0 sec) significantly longer than those of the control (13,8+/-2,5 sec) and laminectomy (11,5+/-2,5 sec - ANOVA, $F_{3,16} = 17.9$, $p < 0.001$, followed by the Duncan test, $p < 0.05$). In the tail-flick test there was no significant difference in the time to flick the tail between control (3,8+/-0,2sec), laminectomy (3,9+/-0,5sec) and contusion (6+/-0sec - ANOVA, $F_{3,16} = 2.12$, $p = 0.13$).

We observed that the hot plate test was the most sensitive behavioral procedure for the evaluation of the lesions. The tail-flick test and the clinical test also showed that injury to spinal cord may occur in the animals during laminectomy, probably produced by the instruments used.

The control group showed no significant expression of Fos protein, whereas expression was present in laminectomy (Figure 1 - upper panel), and contusion group (Figure 1 - lower panel), in the spinal segments located between T10 and L1, corresponding to the site of spinal cord injury, being higher in the group which was submitted to spinal cord injury by contusion (figure 1 and figure 2 - upper panel). The detection of Fos protein in animals submitted only to laminectomy supports the evidence obtained with the behavioral evaluation showing that microtraumas of the spinal cord may occur during surgery. A similar pattern of Fos protein expression was also observed in spinal cord segments above the level of the weight drop injury: cervical and upper thoracic segments. Fos protein was also detected in nuclei of the intermediolateral spinal cord (results not show), in nuclei that regulate the autonomic nervous system in the reticular formation of the brain stem (figure2 - lower panel), the area postrema (Figure 3), the spinal trigeminal caudal nucleus (figure 4 - upper panel) and the nucleus tractus solitaries (figure 2 - middle panel), in animals submitted to spinal cord injury. No significant expression was observed in sham operated rats.

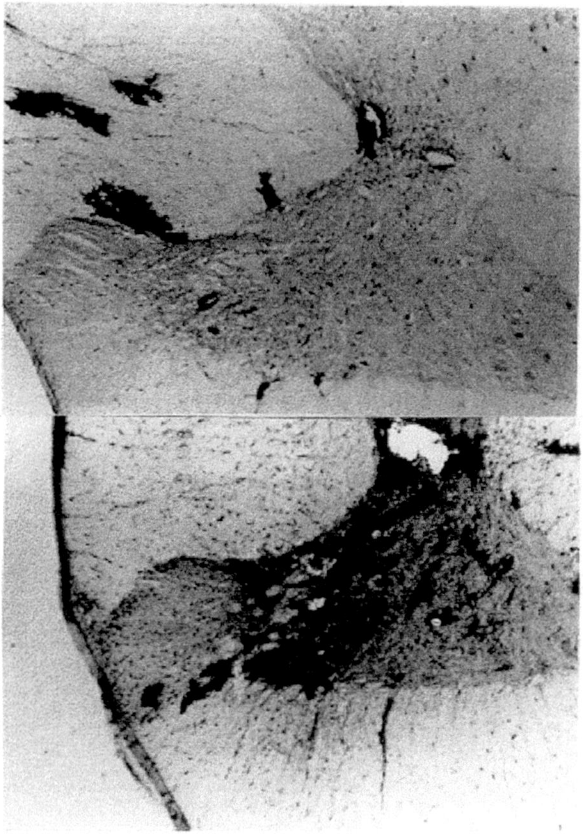

Figure 1. Fos-like immunoreactivity (FLI) in rat spinal cord after laminectomy (upper panel) and weight drop contusion (lower panel. Two hours after the experimental procedure, the rats were anesthetized with i.p. injections of 25% urethane (2.5 ml/kg) and transcardially perfused with saline followed by 4% paraformaldehyde, and the spinal cord was removed for histology and immunocytochemistry. Segments of the cervical, thoracic and lumbar spinal cord and regions of the brain were analyzed, so that histochemical evaluation covered the site of lesion, adjacent spinal segments and brain regions. After being removed, the brains were kept overnight in 30% sucrose solution. Afterwards, 35 μm coronal sections were obtained in a cryostat and processed for Fos immunohistochemistry as described by Del Bel et al. (24). Briefly, tissue sections were successively washed and incubated for 40 h with the primary Fos antibody (1:2000, Cambridge Research Biochemicals).. Sections were then processed by the avidin-biotin immunoperoxidase method (Vectastain ABC kit, Vector Lab.) and FLI was visualized by addition of the chromogen 3,3'-diaminobenzidine (DAB, Sigma) and hydrogen peroxide. The slices were mounted on slides and cover slipped for microscopic observations. FLI could be visualized as a brown reaction product inside neuronal nuclei, whose location was determined using the Paxinos and Watson atlas. (Magnification: 22.5 times).

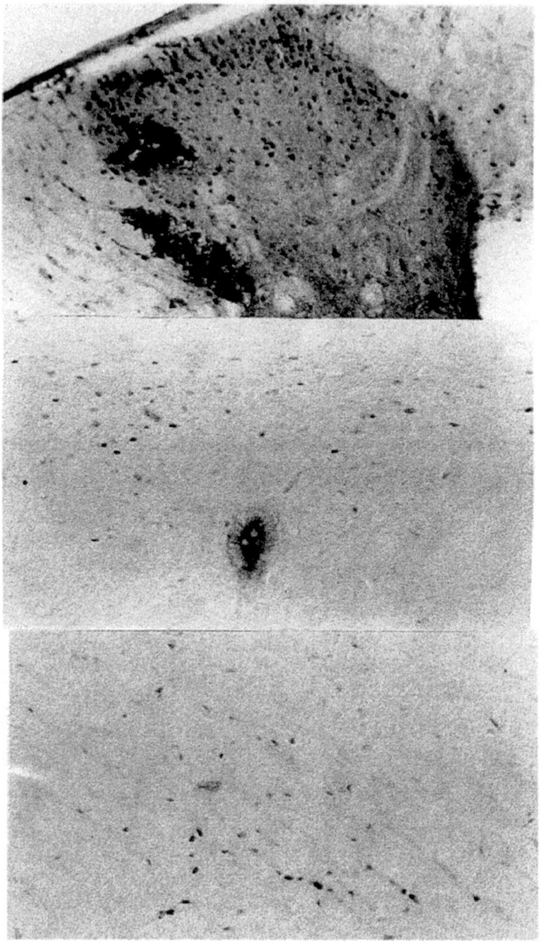

Figure 2. Fos-like immunoreactivity (FLI) in rat spinal cord after weight drop contusion. Upper panel, FLI at dorsal horn; medium panel FLI around central canal; lower panel FLI at neurons of medial mesencephalic reticular formation. (magnification: 56 times)

In the absence of lesions, NADPH-d histochemical labeling of the cervical, lumbar and thoracic spinal cord regions was observed, i.e., the substantia gelatinosa, the intermediate lateral column, cells in the central canal area (Figure 5, 7), and ganglion of dorsal root (Figure 6), observations similar to those reported by Valtschnoff et al. (77), Dun et al., (26), Aimi et al. (1) and Saito et al. (61). Following weight contusion NADPH-d staining was evident in cytoplasmic region of neuronal soma, as well as in dendrites in ventral horn neurons (Figure 7, 8, 9 and 10); no staining of the cells was evident before lesion (figure 9 and 10). It was evident from horizontal sections that some intermediolateral nucleus neurons send extend processes into the white matter (Figure 9). An increase in the number of cells positively stained for NADPH-d activity was observed in cells of the intermediolateral spinal cord of lamina 5 in contusion group (figure 7 and 8).

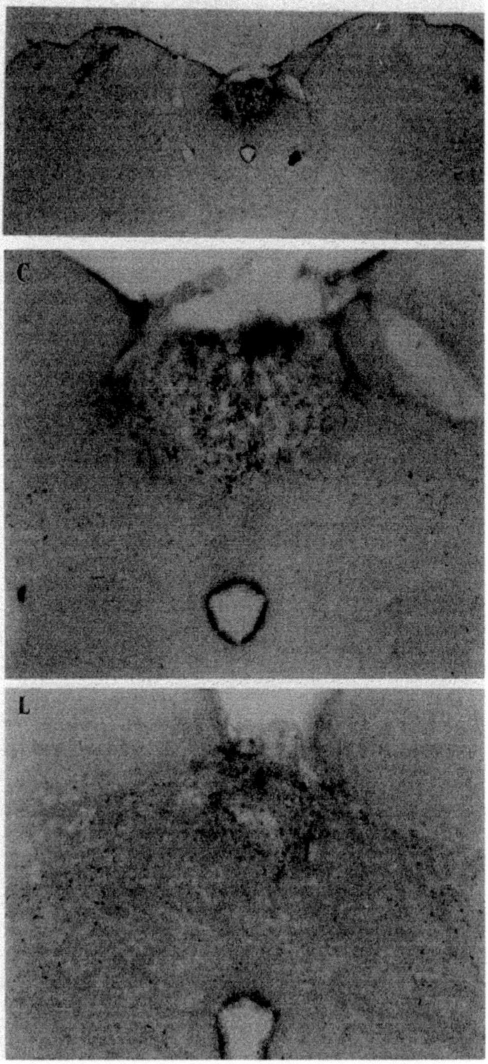

Figure 3. Fos-like immunoreactivity (FLI) in rat spinal cord after weight drop contusion. Upper panel, FLI at area postrema and around central canal (C - laminectomy group- low magnification); middle panel FLI at area postrema and central canal (laminectomy, 56 times amplification); lower panel FLI at neurons of reticular formation and around central canal, L - contusion group (56 times amplification).

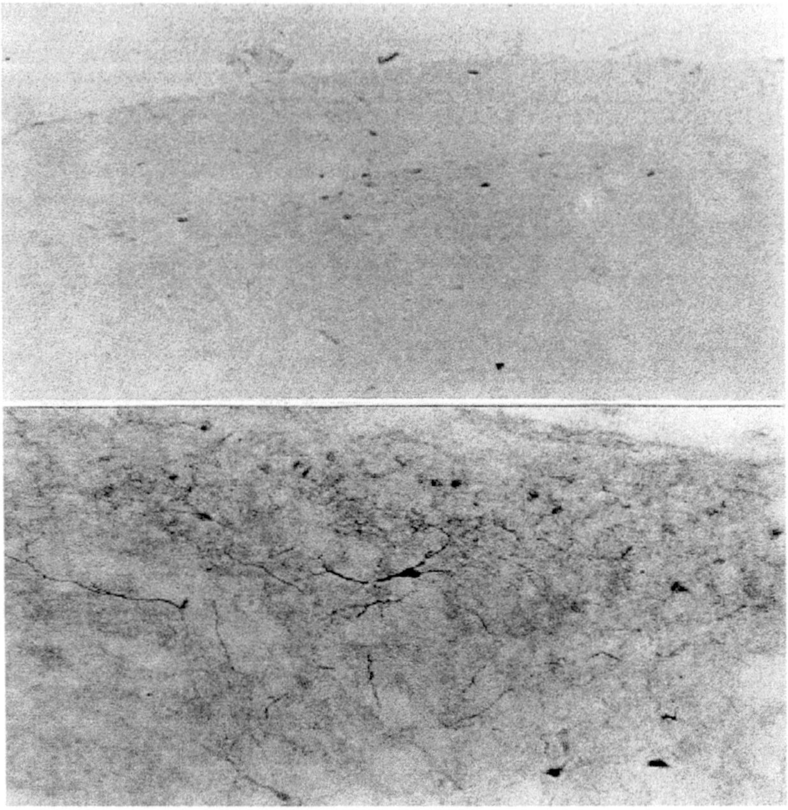

Figure 4. Fos-like immunoreactivity (FLI) in rat spinal cord after weight drop contusion. Upper panel, FLI at nucleus trigeminalis caudalis; lower panel NADPH-diaphorase/NOS positive cells. The NADPH diaphorase histochemical reaction was performed using the method employed by Vincent and Kimura (79). Rats were anesthetized i.p. with 25% urethane and perfused through the left ventricle with 100 ml of 0.9% saline followed by 200 ml 4% paraformaldehyde, and 0.1 M phosphate buffer solution. The brains were removed, soaked for 2 hrs. in the 4% paraformaldehyde, 0.1 M phosphate buffer solution and overnight in 30% saccharose 0.1 M phosphate buffer solution, frozen in liquid nitrogen, and then stored at -70°C. Twelve µm sections were then cut with a cryostat. NADPH-diaphorase activity was demonstrated by incubating sections in 0.1 M phosphate buffer, pH 7.4, containing 0.3% Triton-X100, 0.1 mg/ml nitrobluetetrazolium and 1.0 mg/ml ß-NADPH at 37°C for 30-60 min in the dark. The sections were then dipped for 10 min in phosphate buffer, dried, rinsed in distilled water, dried again and mounted for microscopic observation. NOS activity was demonstrated by NADPH-d. The blue product of this reaction showed a distribution in the cytoplasm, dendrites and axons of neurons. Upper panel left and right, central canal; lower panel left, dorsal horn NADPH diaphorase neurons; lower panel right: neurons from intermedio lateral column (amplification 56 times).

The neurons surrounding the central canal in lamina X show exceptionally high levels of NADPH-diaphorase activity (figure 5 and 8). The region of spinal cord around the central canal is particularly susceptible to post-traumatic necrosis and receive visceral afferent (35). Potential visceral regulatory roles for NOS in these cells appear to be similar to those of the intermediolateral region.

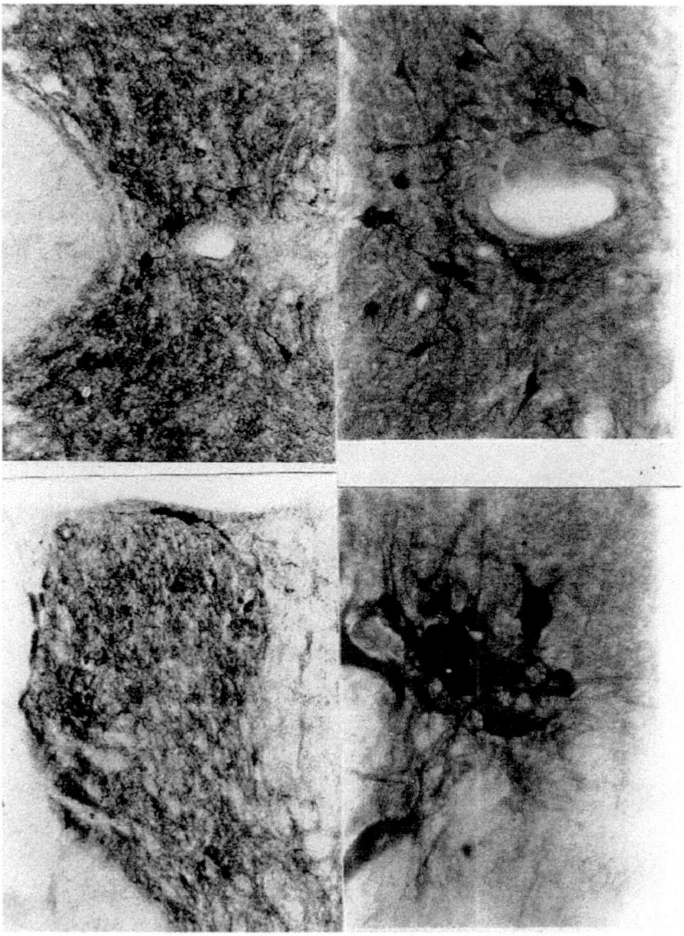

Figure 5: NADPH-diaphorase/NOS positive cells located in rat spinal cord Upper panel NADPH-d positive cells around central canal left - low magnification, 22,5 times, and right - higher magnification, 56 times; lower panel left, dorsal horn NADPH diaphorase neurons; lower panel right: higher magnification (56 times) showing several NADPH-d positive cells in the dorsal horn superficial layers; left: higher magnification (56 times) illustrating several neurons NADPH-d positive, in the intermediolateral cell column from intermedio lateral column.

NADPH-d neurons are scattered in the dorsal column nuclei and a wide variety of reticular nuclei, such as the rostral ventrolateral reticular nucleus (results not shown). Among the lower brain stem centers involved in central autonomic control, NADPH-d neurons occur in the dorsal motor nucleus of the vagus, the nucleus ambiguous, the nucleus tractus solitaries and the ventrolateral medulla oblongata (53).

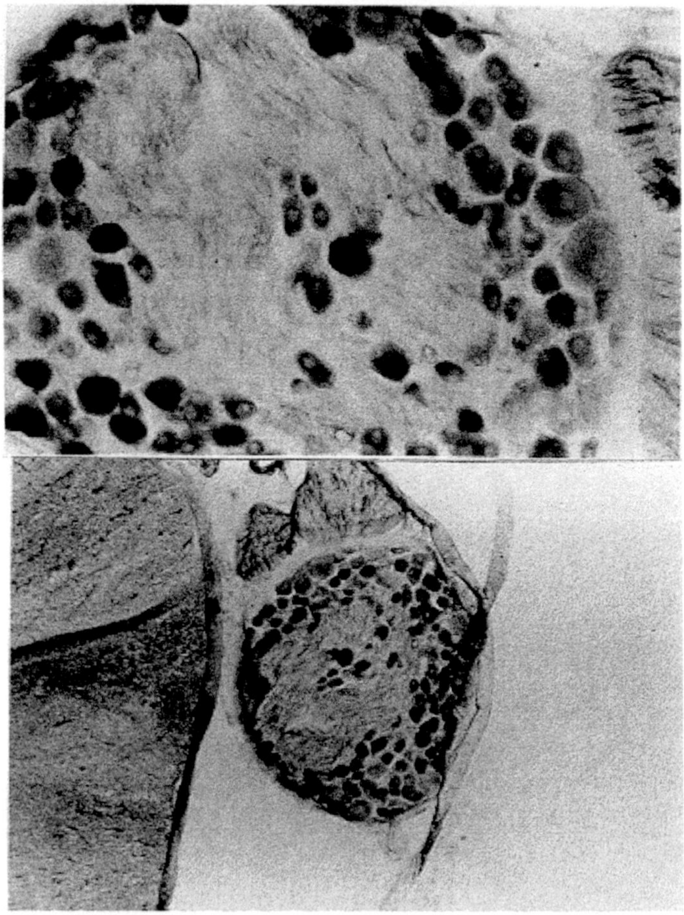

Figure 6: NADPH-diaphorase positive cells located in rat spinal cord: upper panel NADPH diaphorase neurons in dorsal root ganglion cells, higher magnification (56 times); lower panel low magnification.

Moreover, Fos protein expression in the region corresponding to lamina V of the spinal column of the rat in neurons around the central canal of the spinal cord, in superficial laminae I and II of the dorsal horn (Figure 2), and in cells of the intermediolateral column parallels the increase in number of NADPH diaphorase-positive cells in the same region in injured animals (Figure 9). In contrast, there was no Fos protein expression at the ventral root motoneurons, despite an increase in NADPH-diaphorase activity.

The location of IEG proteins in neurons may precisely reflect the fact that the cell was somehow stimulated and the distribution of the protein may be specifically related to the type of stimulus and to the physiological status of the animal. This fact was demonstrated in the response of the spinal column to noxious stimulation such as pain (38, 40, 70). Thus, IEG expression should not simply be evaluated as a marker of activation, but as an indicator of the occurrence of environmental stimulation, which requires long-term variations in certain aspects of neuronal physiology. Munglani and Hunt (56) reported that c-Jun induction in the spinal cord after peripheral injury may be associated with regeneration processes. Raghupathi et al. (59), in turn, demonstrated that the induction of proto-oncogenes such as c-Fos and c-Jun may precede the cell death of spinal neurons maintained in culture after chemical injury. Morris, (53) showed that nitric oxide stimulated IEG expression in striatal neurons. Lee et al (44) showed that nitric oxide mediates Fos expression in the spinal cord induced by mechanical noxious stimulation.

Detection of FOS positive cells at regions far away from the lesion site corroborate evidence of subclinical brain influence in clinically complete spinal cord injury (62). Krenz and Waver (42) suggested that the increased response to sensory stimulation below the level of spinal cord injury may result in an increase in muscle tonus and aberrant movements, elevation of blood pressure (33) accompanied by headache, convulsive seizures and even a cerebrovascular accident. According to these investigators, these alterations does not result exclusively from the absence of descending inhibition of motor and autonomic neurons, but may also be influenced by superior nervous centers. The nuclei of the autonomic nervous system of the reticular formation in which Fos protein was detected after impact injury are involved in regulation of blood pressure, which is altered in tetraplegic and paraplegic patients as well as after experimental spinal cord injury in rats (22, 25, 33). Cells of the intermediolateral column represent preganglionic sympathetic fibers that emerge via the ventral roots and project to various sympathetic ganglia (57).

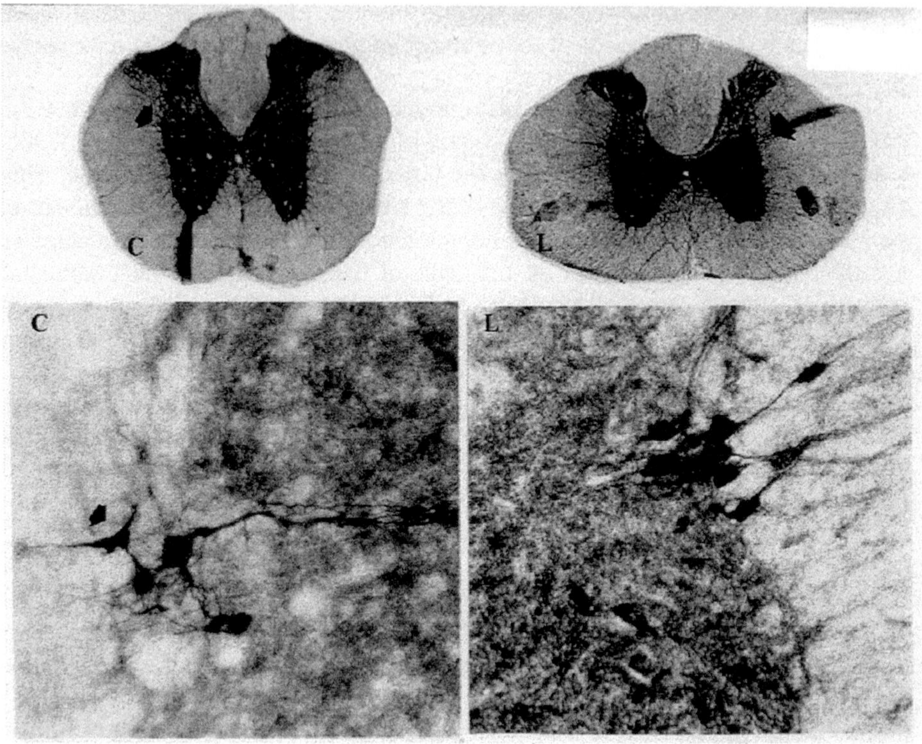

Figure 7: NADPH-diaphorase positive cells located in rat spinal cord Upper panel left: 4 times magnification of transverse section of thoracic spinal cord (C= control); lower panel left, 56 times magnification of NADPH-d positive cells in the intermediolateral column; upper panel right, 4 times magnification of transverse section of thoracic spinal cord (L= lesion); lower panel right: 56 times magnification of NADPH-d positive cells in the intermediolateral column; from intermedio lateral column.

Traumatic injury of the spinal cord may have severe consequences for the patients and lesions of the spinal cord may be increased by the neuronal death processes occurring around the site of lesion (25). These cell alterations may also occur in neuronal cells locates in nuclei far away from the site of injury, expanding the process of spinal cord lesion (62). Traumatic or ischemic injury of the CNS triggers reactive biochemical variations, some of them being destructive and others neuroprotective (22). With respect to the functional relevance of the induction of the c-Fos proto-oncogene and of NADPH diaphorase-positive neurons after injury, Dawson and Dawson, (19) suggest that since the cell cannot sequester and regulate the local concentration of NO, the key to regulating NO activity is to control NO synthesis. Our results suggest is suggest that the increased NADPH-d activity of the damaged neurons may be involved in neural processes such as regeneration and/or degeneration, leading to neuronal survival and/or death. Morgan and Curran (50-52) and Hokfelt et al. (37) have suggested that proteins may be synthesized in response to injury and may participate in the neuronal response secondary to injury (67, 69, 70, 71). Although neuronal NOS is constitutively expressed, under some pathologic insults NOS can be induced in certain cells through new protein synthesis (74).

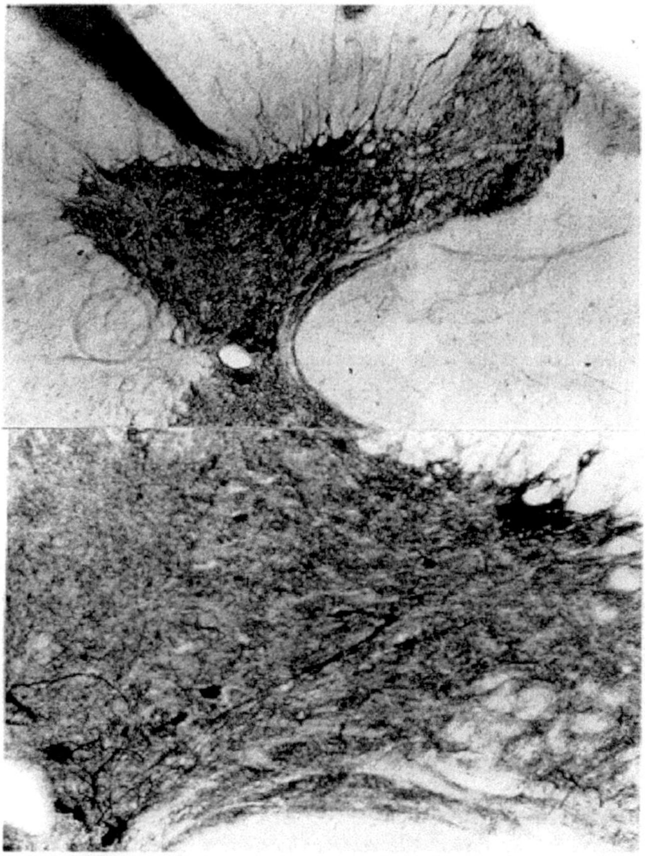

Figure 8: NADPH-diaphorase positive cells located in rat thoracic spinal cord Upper panel, low magnification (22,5 times) central canal, intermediolateral column, dorsal horn superficial layers; lower panel NADPH diaphorase neurons in intermediolateral column extending process to central canal neurons (56 times).

NOS is induced in motor neurons of the spinal cord after avulsion or lesion of C6 or C7 root (74-76) and in intermediolateral lumbosacral neurons (IML-2) neurons after pelvic axotomy (3, 7, 78, 80). The expression of NOS seems to indicate the subsequent death of motoneurons (74-76). Consistent with the concept , treatment with NOS inhibitor dramatically increases the number of surviving neurons (75). In addition, recent studies have reported that NO may modulate gene transcription and translation.

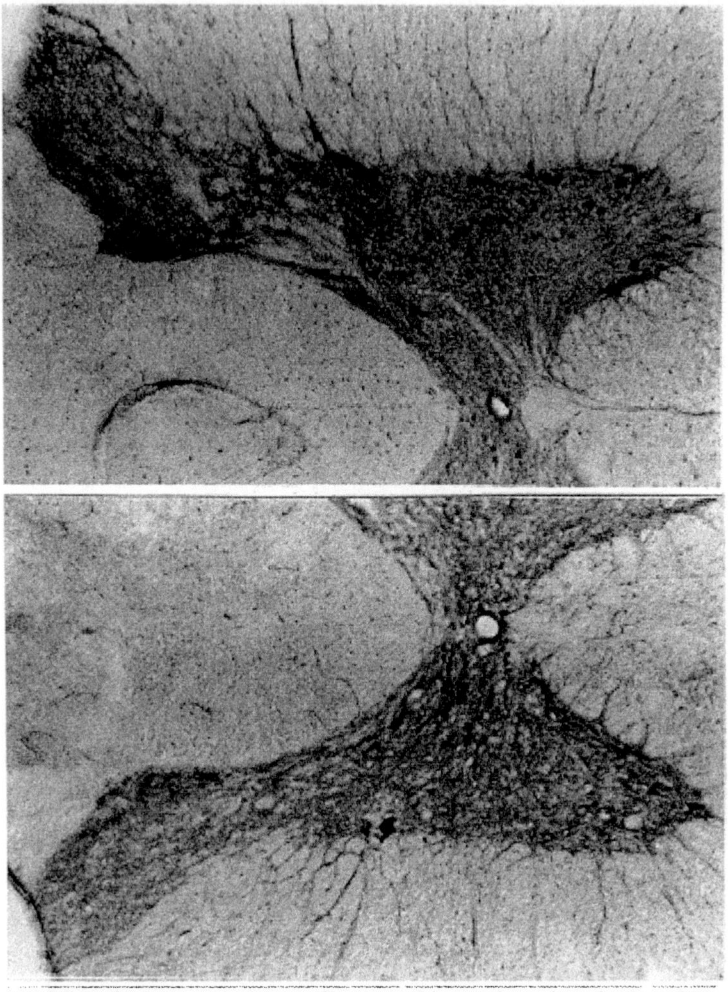

Figure 9: NADPH-diaphorase positive cells located in rat thoracic spinal cord Upper panel lesioned, central canal, intermediolateral cells and ventral horn neurons, absent at control ventral horn at lower panel left (22,5 times amplification).

Thus, in NO-positive neurons NOS may potentiate the effects of calcium and stimulate the expression of genes such as c-Fos (28, 30, 32, 44). Increased NADPH-d levels in spinal cord neurons and white matter may suggest effects on spinal neurons by release of NO from presynaptic terminals (46).

A beneficial effect of NOS is also indicated by co-expression with c-Jun in NADPH diaphorase labeled neurons (28). In contrast, data from intratecal infusion of Nitric Oxide synthase inhibitor L-NOARG after experimental spinal cord injury in guinea pigs do not support the hypotheses that local NO production plays a significant role in secondary pathology of injury in the compression injury model (15). NO activates glial cells of the dorsal root ganglia by stimulating the cGMP system (1, 3, 49), and activation of glial cells plays an important trophic role in lesioned neurons (17). Up-regulation of NADPH-d/NOS and a presumable increase in NO release could enhance regeneration by increasing blood flow (53).

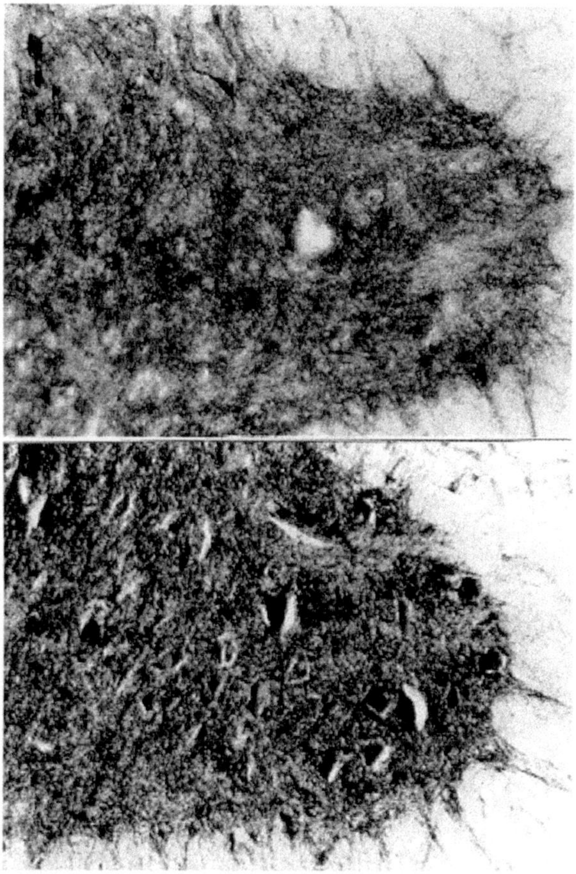

Figure 10: NADPH-diaphorase positive cells located in rat thoracic spinal cord Upper panel: ventral horn neurons, control lower panel ventral horn NADPH-d positive neurons at lesioned animals (56 times amplification).

The regenerating ability of the CNS of mammals is limited, with affected neural pathways often being unable to recover. One of the major challenges in this field is to recover neuron activity, with reestablishment of he physiological function of damaged neurons (58, 63). Advances in clinical and basic research have led to a better understanding of the mechanism of injury and this information is expected to be effectively applied to treatment of the damaged area. The initial reports cited here showed some changes in neuronal phenotype and gene expression after spinal cord injuries. We believe that the knowledge of the temporal of neuronal activation/degeneration process may offer a temporal window for therapeutic intervention.

Acknowledgments: We are indebted to Dr. Francisco S. Guimarães and Dr. Wilian Alves do Prado, Department of Pharmacology, FMRP, USP, for useful discussions; to José Carlos de Aguiar, Eleni L. T. Gomes, Paulo Castania, Célia Aparecida da Silva e Renata Caldo for excellent technical assistance. Special thanks to Dr. Luiz G. Brentegani, Department of Pathology, FORP, USP, and Hermano T Machado for photographic assistance. We thank Dr Guimarães for his critical review of the manuscript. Part of this work represents Dr. Clovis A. Borges MSc Thesis at the Department of Orthopedic Surgery. Financial support: FAPESP, CNPQ and FAEPA.

References

1. Aimi, Y. et al.: Localization of NADPH-Diaphorase-containing neurons in sensory ganglia of rat. J.Comp. Neurol. 306: 382-392 (1992)
2. Albin, MS et al.: Study of functional recovery produced by delayed cooling after spinal cord injury in primates. J Neurosurg. 29:113 (1969)
3. Ando, A et al: Changes of NADPH Diaphorase activity in the lumbosacral intermediolateral neurons of the rat after pelvic axotomy Brain Research Bulletin 40: 37-42 (1996)
4. Allen, A.R: Surgery of experimental lesion of spinal cord equivalent to crush injury of fracture dislocation of spinal column. A preliminary report. Jama 57:878-880 (1911)
5. Allen, AR: Remarks on the histopathological changes in the spinal cord due to impact: an experimental study, J Nerv. Ment. Dis. 41:141 (1914)
6. Bartholdi, D; Schawb, M.E.: Methylprednisolone inhibits early inflammatory processes but not ischemic cell death after experimental spinal cord lesion in the rat. Brain-Res, 672:177-86 (1995).
7. Blottner, D.; Baungarten, H.G.: Nitric oxide synthase (NOS)-containing sympathoadrenal cholinergic neurons of the rat IML-cell column: evidence from histochemistry, immunohistochemistry and retrograde labeling. J. Comp. Neurol. 316:45-55 (1992)
8. Blottner, D et al.: Hystochemistry of NOS in the nervous system. Histochem. J. 27: 785-811 (1995)
9. Bolman, H.H. et al.: Spinal cord monitoring of experimental incomplete spinal cord injury: A preliminary report. Spine 6:428-436 (1981)
10. Compston, A.: Brain repair: an overview. J. Neurol., 241: S1-S4 (1994)
11. Cheng, H. et al.: Spinal cord repair in adult paraplegic rats: partial restoration of hind limb function. Science 273:510-513 (1996)
12. Choi, D.W.: Glutamate neurotoxity and disease of the nervous system, Neuron 1, 623-634 (1988)
13. Choi, D.W.: Excitotoxic cell death. J. Neurobiol. 23,1261-1276 (1992)
14. Chopp, M.: The role of heat shock proteins and immediate early genes in central nervous system normal function and pathology Curr. Opin. Neurol. Neurosurg. 6,6-10 (1993)

15. Coehn, T.I. et al.: Intratecal Infusion of Nitric Oxide Synthase Inhibitor L-NOARG after experimental spinal cord injury in guinea pigs J. Neurotr. 13: 361-365 (1996)
16. Dawson T.M. et al.: A novel neuronal messenger molecule in brain: the free radical Nitric oxide Ann. Neurol. 32,297-311 (1992)
17. Dawson, T.M. et al.: Nitric oxide synthase and neuronal NADPH diaphorase are identical in brain and peripheral tissue. Proc. Natl. Acad. Sci. **88**, 7797-7801 (1991)
18. Dawson, T.M. et al. Neuroprotective effects of gangliosides may involve inhibition of nitric oxide synthase. Ann. Neurol. 37, 115-118 (1995)
19. Dawson, V.L. and Dawson, T.M.: Nitric Oxide actions in neurochemistry. Neurochem. Int. 29:97-110 (1996)
20. Del Bel, E.A.et al.: Induction of the c-fos proto-oncogene in the rat pineal gland during stress. Braz. J. Med. Biol. Res. 26:975-981 (1993)
21. Del Bel, E.A. et al.: Anticonvulsant and proconvulsant rolos of Nitric Oxide in experimental epilepsy models Brazilian J. Med. Biol. Res., in press
22. De Felipe, C. et al.: The role of immediate early genes in the regeneration of the central nervous system. Adv. Neurol. 59, 263-271 (1993)
23. Del Bel, E.A. Investigation of neuronal functional activation under stress. International Brain Research Organization News 24(3): 3-4, 1996.
24. Del Bel, E.A. et al.: C-FOS mRNA and protein Distribution patterns in rat brain after restraint stress and pentylenetetrazol-induced seizures Cellular and Molecular Neurobiology (1997)
25. Dursart, I., Schwab, M. E.: Secondary cell death and the spinal cord. Europ. J. Neuros. 6:712-724 (1994)
26. Dun, N.J.., et al.: Nitric oxide synthase immunoreactivity in rat spinal cord Neuroscience Letters, 147, 217-220 (1992)
27. Faden A I.: Experimental neurobiology of central nervous system trauma. Crit. Ver. Neurobiol. 7:175-186 (1993)
28. Fiallos-Estrada, C.E. et al.: Long-lasting increase of NOS immunoreactivity, NADPH-diaphorase reaction and c-Jun co-expression in rat dorsal root ganglion neurons following sciatic nerve transection. Lett., 150: 169-173 (1993)
29. Filkestein, S.D. et al.: Experimental spinal cord injury: qualitative and quantitative histopathologic evaluation J. Neurotrauma 7: 29-40 (1990)
30. Finkbeiner, S.: Calcium-mediated gene expression: mechanism for neuronal plasticity and survival. Neurocientist 6: 317-320 (1995)
31. Garthwait, J.: Glutamate, nitric oxide and cell-cell signaling in the nervous system. TrendsNeurosc. 14:60-68 (1991)
32. Gonzalez, M.F. et al.: Axotomy increases NADPH-diaphorase staining in vagal motor neurons. Brain Res. Bull. 18: 417-427 (1987)
33. Guha, A, et al.: Spinal cord blood flow and systemic blood pressure after experimental spinal cord injury in rats. Stroke 3: 372-377 (1989)
34. Hapel R.D. et al.: Ca2+ accumulation in experimental spinal cord trauma Brain Res. 211: 476-479 (1981)

35. Herdegen, T. et al.: The transcription factors c-jun, jun D and CREB proteins, but not FOS and KROX-24, are differentially regulated in axotomized neurons following transsection of rat sciatic nerve. Mol. Brain Res. 14: 155-165 (1992)
36. Hughes, P. , Dragunow, M.: Muscarinic receptor-mediated induction of Fos protein in rat brain. Neurosc. Lett. 150:122-126 (1993)
37. Hökfelt T. et al.: Messenger plasticity in primary sensory neurons following axotomy and its functional implications Trends in Neurosci.17: 22-30 (1994)
38. Hunt, S.P.et al.: Induction of c-fos-like protein in spinal cord 328:632-634 (1987)
39. Iadecola, C: Brigth and dark sides of nitric oxide in ischemic brain injury, TINS 20, 132-139 (1997)
40. Jenkins, R., S.P.Hunt: Long term increase in levels of c-jun mRNA and Jun Protein-like immunoreactivity in motor and sensory neurons following axon damage, Neurosci. Lett., 129, 107-110 (1991)
41. Jenkins, R. et al.: Differential expression of immmediate early genes in rubrospinal neurons following axotomy in rat. Europ. J. Neuros. 5: 203-209 (1993)
42. Krenz N.R., Weaver L.C.: Sprouting of primary afferent fibers after spinal cord transection Abstracts of Society for Neuroscience 26th Annual Meeting Washington D.C. part 2;1488 (1996)
43. Liu, D. et al.: Excitatory aminoacids rise to toxic levels upon impact injury to the rat spinal cord. Brain Res. 547: 344-348 (1991)
44. Lee, J.H. et al: Nitric Oxide mediates Fos expression in the spinal cord induced by mechanical noxious stimulation, NeuroReport 3, 841-844 (1992)
45. Lino de Oliveira C. et al.: c-jun mRNA expression in the hippocampal formation induced by stress Brain Research (1997)
46. McElhaney, M.R. et al.: Astrocytes but not microglia express NADPH-diaphorase activity after motor neuron injury in the rat Neuroscience Letters 180:67-70 (1994)
47. Meldrum B., Garthwaite, J.: Excitatory amino acid neurotoxity and neurodegenerative disease. Trends Pharmacol Sci 11: 379-387 (1990)
48. Moncada, S., Higgs, A.: The l-arginine pathway. N. Engl. J. Med. 329, 2002-2012 (1993)
49. Moncada, S. et al.: Nitric oxide: physiology, pathophysiology, and pharmacology. Pharmacol. Rev. 43, 109-142 (1991)
50. Morgan, J.I. et al: Mapping patterns of c-fos expression in the CNS after seizure. Science 237:192-197 (1987)
51. Morgan, J.I., Curran, T.: Proto-oncogenes: beyond second messengers. In: Bloom, F.E. and Kupfer, D.J. (eds.). Psuchopharmacology: the forth generation of progress. Raven Press, New York, pp. 631-642 (1995)
52. Morgan, J.I., Currant,T.: The immediate-early gene response and neural death and regeneration. Neuroscientist 1: 68-75 (1995)
53. Morris, B.J.: Stimulation of immediate early gene expression in striatal neurons by nitric oxide. J. Biol. Hem 270: 24740-24744 (1995)

54. Morris, R. et al.. Nitric oxide may act as a messenger between dorsal root ganglion neurons and their satellite cells. Neurotics. Lett., 137: 29-32.
55. Morris, B, J.. Stimulation of immediate early gene expression in striatal neurons by nitric oxide J. Biol. Chem. 270,:24740-24744 (1995)
56. Munglani, R. Hunt, S.P.: Proto-oncogenes: basic concepts and stimulation induced changes in spinal cord. In: Wiesenfeld-Hallin,S; Sharma,N; Nyberg,F: eds. Neuropeptides in the Spinal Cord Fundamental and clinical Aspects (1995)
57. Nadelhaf I., Booth, A M.: The location and morphology of preganglionic neurons and the distribution of visceral afferents from the rat pelvic nerve: a horseradish peroxidase study J. Comp. Neurol 226: 238-245 (1984)
58. Nicholls, J, Saunders, N.: Regeneration of immature mammalian spinal cord after injury. Tins 19(6) 229-234 (1996)
59. Rhagupathy L.J. et al.: Increased immunoreactivity of c-jun and jun-B but not jun-D in dying and dead neurons after physical injury Abstracts of Society for Neuroscience 26th Annual Meeting Washington D.C. part 1:21 (1996).
60. Rothmam, S.M., Olney, J.W.: Excitotoxicity and the NMDAA receptor. Trends Neurosci. 10: 299-302 (1987)
61. Saito, S. et al.: Rat spinal cord neurons contain NOS. Neuroscience; 59(2): 447-56 (1994)
62. Sherwood A M., et al.: Evidence of subclinical brain influence in clinically complete spinal cord injury: discomplete SCI J. Neurol. Sci. 110: 90-98 (1992)
63. Schwab, M.E., Bartholdi, D.: Degeneration and regeneration of axons in the lesioned spinal cord. Physiological Reviews 76:319-370 (1996)
64. Smeyne, R.J. et al.: Continuous c-fos expression precedes programmed cell death in vivo Letters to Nature. 363:166-169 (1993)
65. Sun, F.Y., Faden, A: High- and low-affinity NMDA receptor binding sites in rat spinal cord: effects: of traumatic injury. Brain Research. 666: 88-92 (1994)
66. Tatter, S.B., et al.: Neurotrophic factor protection against excitotoxic neuronal death The Neurocientist 5: 286-297 (1995)
67. Theriault , E., et al.: Elevated gene expression in the red nucleus after spinal cord compression injury. NeuroReport 3: 559-562 (1992)
68. Titze-de-Almeida, R., et al.: Midazolam and the N-methyl-D-aspartate (NMDA) receptor antagonist 2-amino-7-phosphonoheptanoic acid (AP-7) attenuate stress-induced expression of c-fos mRNA in the dentate gyrus. Cell. Mol. Neurobioloy 14:373-380 (1994)
69. Titze-de-Almeida, R., et al.: Stress-induced expression of c-fos proto-oncogene in the hippocampal formation. Braz. J. Med. Biol. Res. 27:1083-1088 (1994)
70. Tolle, T.R., et al.: Application of morphine prior to noxious stimulation differentially modulates expression of fos, jun and Krox-24 proteins in rat spinal cord neurons. Neuroscience. Jan;58(2): 305-21 (1994).
71. Traub, R.J., et al.: NADPH-diaphorase histochemistry provides evidence for a bilateral, somatatopically inappropriate response to unilateral hindpaw inflammation in the rat. Brain-Res.647:113-23 (1994)

72. Wan-Hua, A Y.: nitric oxide in motor neurons after axotomy The J. Histoch. Cytochem. 42:451-457 (1994)
73. Wrathll, J.R.: Spinal cord injury models J. Nerotr. 9, S129-S134 (1992)
74. Wu, W.. (1993) Expression of nitric-oxide synthase (NOS) in injured CNS neurons as shown by NADPH diaphorase histochemistry. Exp-Neurl. 120(2):153-9 (1994)
75. Wu, W., Li, L: Inhibition of NOS reduces motoneuron death due to spinal root. Neurosci-Lett. Apr 30:153(2):121-4 (1994
76. Wu, W. et al.: Neuronal NOS is induced in spinal neurons by traumatic injury Neuroscience 61:719-726 (1994)
77. Valtschanoff, J.G., et al.: NADPH diaphorase in spinal cord of rats The Journal Comp. Neurol. 321:209-222 (1992)
78. Verge, V., et al.. Marked Increase in Nitric Oxide synthase mRNA in dorsal root ganglia after peripheral axotomy: in sit hybridization and functional studies. Proc. Natl. Acad. Sci. USA 89: 11617-11621 (1992)
79. Vicent, S.R., Kimura, H.: Histochemical mapping of NOS in the rat brain. Neurosci, 46:755-784 (1992)
80. Vizzard, M., et al.: The effect of rhizotomy on NADPH diaphorase staining in the lumbar spinal cord of the rat. Brain-Res. 2;1-:349-353 (1995)
81. Young, W., Koreh, I.: Potassium and calcium changes in injured spinal cords. Brain Res. 365: 42-53(1986)

Fourier analysis applied to the study of the electrical activity of midbrain dopaminergic neurons

Michele Di Mascio, Giuseppe Di Giovanni and Ennio Esposito

Istituto di Ricerche Farmacologiche "Mario Negri", Consorzio "Mario Negri Sud", 66030 Santa Maria Imbaro (Chieti), Italy

Abstract. In this study, electrophysiological techniques and computational methods were used to investigate the effect of the selective serotonin reuptake inhibitors fluvoxamine, paroxetine, sertraline and citalopram on the basal activity of dopamine (DA) neurons in the ventral tegmental area. Acute injection of fluvoxamine, paroxetine, sertraline and citalopram (20-1280 μg/kg, i.v) caused a dose-dependent inhibition of some ventral tegmental area DA neurons but it did not affect the basal firing rate of other DA cells. A Fast-Fourier-Transformation based analysis of the basal activity of 45 ventral tegmental area DA neurons showed a positive correlation between the value of a functional operator (Ψ) equivalent to the density-power-spectrum of the signals and the degree of selective serotonin reuptake inhibitors-induced inhibition of ventral tegmental area DA cells. All ventral tegmental area DA neurons sampled were subdivided into two subclasses: (A) neurons with no changes in their basal firing rate and (B) neurons showing an approximately linear inhibitory effect in response to selective serotonin reuptake inhibitors. The neurons belonging to the subclass A showed a more regular behavior of the interspike interval functions corresponding to lower values detected by the functional operator Ψ, whereas the neurons belonging to the subclass B showed a less regular behavior of interspike interval functions corresponding to higher Ψ values detected by the same functional operator. SSRIs also caused a dose-dependent increase of the percentage of spikes occurring in bursts in neurons belonging to subclass A (low values of Ψ), whereas the mean basal firing rate of these cells was not affected. Moreover, administration of the 5-HT_{1A} receptor agonist 8-hydroxy-2-(di-*n*-propylamino)tetralin (8-OH-DPAT) (1.25-80 μg/kg, i.v) increased the firing rate of midbrain DA neurons with higher Ψ values. It is suggested that this difference in density-power-spectrum could reflect the asymmetry of serotonergic input to the ventral tegmental area DA neurons, and the differential effects of selective serotonin reuptake inhibitors and of 8-OH-DPAT on these neurons might depend on the characteristics of their basal firing mode.

Key-words. Antidepressant drugs, electrophysiology, single cell recording, burst analysis, Fast-Fourier-Transformation.

1 Introduction

There is extensive evidence that the activity of midbrain dopamine (DA)-containing neurons is modulated by the serotonergic system. Neuroanatomical studies have demonstrated that both the substantia nigra, *pars compacta* (SNc) and the ventral tegmental area (VTA) receive afferent projections from 5-hydroxytryptamine (5-HT)-containing axon terminals originating in the midbrain raphe nuclei [1,12,17]. Electrophysiological experiments have shown that 5-HT exerts an inhibitory effect on the activity of DA neurons in the VTA [16]. Thus, 8-hydroxy-2-(di-*n*-proprylamino)tetralin (8-OH-DPAT), a prototypical 5-HT_{1A} receptor agonist, which potently inhibits 5-HT neurons in both dorsal and median raphe nuclei increased the mean basal firing rate of DA cells in both the SNc and the VTA [10,16]. Mixed 5-$HT_{2C/2B}$ receptor agonists, such as trifluoromethylphenylpiperazine (TFMPP) and m-chlorophenylpiperazine (mCPP), significantly reduced the activity of DA neurons in the VTA and weakly inhibited the SNc DA cells [10,16]. These findings indicate that 5-HT may exert a tonic inhibitory influence upon the activity of midbrain DA neurons by acting through 5-$HT_{2C/2B}$ receptor subtypes. Consistent with these data, it has recently been found that acute administration of selective 5-HT reuptake inhibitors such as fluoxetine and citalopram reduced the basal firing rate of VTA DA cells by increasing the serotonergic transmission [15]. In the latter study it was found that the response to citalopram was less pronounced than that to fluoxetine and was also variable, as it clearly inhibited the majority (12 of 16) of DA neurons tested, whereas it did not affect the activity of other DA cells (4 of 16) [15]. The differential effect of the drug did not depend on the basal firing rate of the neurons sampled. Moreover, fluoxetine and citalopram did not induce any changes in the firing pattern of the VTA DA neurons sampled, expressed as the percentage of spikes occurring in bursts [15].

In this study, the effect of four selective inhibitors of 5-HT uptake (fluvoxamine, paroxetine, sertraline and citalopram) and of a 5-HT_{1A} agonist (8-OH-DPAT) on the activity of VTA DA neurons was investigated, with the aim of identifying a physical parameter associated with the firing pattern of DA neurons which could be correlated with the effect of the drugs.

2 Material and methods

2.1 Animals

Male Sprague-Dawley rats (Consorzio Mario Negri Sud, Italy) weighing 300 to 350 g were used. Animals were kept at constant room temperature (21 ± 2°C) and relative humidity (60%) with a 12-h light-dark cycle (dark from 8 p.m.) and had free access to water and food. Procedures involving animals and their care were conducted in conformity with the institutional guidelines that are in compliance with national (D.L. n. 116, G.U., suppl. 40, 18 Febbraio 1992) and international laws and policies (*EEC Council Directive* 86/609, OJ L 358,1, Dec. 12, 1987; *NIH Guide for the Care and Use of Laboratory Animals*, NIH Publication N. 85-23, 1985 and Guidelines for the Use of Animals in Biomedical Research, Thromb. Haemost. **58**, 1078-1084, 1987).

2.2 Surgical and recording procedures

Rats were anesthetized with chloral hydrate (400 mg/kg, i.p.) and mounted on a stereotaxic instrument (SR-6, Narishige, Japan). Supplemental doses of anesthetic were administered *via* a lateral tail vein cannula. Throughout the experiment the animal's body temperature was maintained at 36-37° C by a thermostatically regulated heating pad. The anteroposterior coordinates, relative to the interaural line, for placement of the recording electrode in the VTA were: anterior 2.7 to 3.4 mm, lateral 0.3 to 0.5 mm; the vertical coordinate was: 7 to 8 mm ventral to the level of exposed tissue [14]. Extracellular recordings were performed with single-barrel micropipettes (4-7 MΩ resistance containing 2% pontamine sky blue dye in 2 M NaCl). Dopaminergic neurons were identified by their location, waveform, firing rate and pattern [2,4,18]. Electrical signals of spike activity were passed through a high impedance amplifier whose output was led into an analog oscilloscope, audio monitor and window discriminator. Unit activity was then converted to an integrated histogram by a rate-averaging computer and displayed as spikes per 10-sec intervals. It has been shown that these neurons have two modes of firing: single-spiking and bursting [5,6]. Since the refractory period of DA cells is approximately 10 ms [3,20], data files with interspike intervals shorter than 10 ms were not further analysed.

After each experiment, the recording site was marked by the ejection of pontamine sky blue dye from the electrode using a -20 μA current for 10 min. Brains were removed and placed in 10% buffered formalin for 2 days before histological examination. Frozen sections were cut at 40-μm intervals and stained with neutral red. Microscopic examination of the sections was carried out to verify that the electrode tip was in the VTA.

2.3 Drug administration protocols

All the drugs were administered i.v. (*via* a lateral tail vein) in exponentially increasing doses every 2 min, and the effect on the activity of dopaminergic neurons was recorded. Only one cell per animal was studied. Fluvoxamine (20-1280 μg/kg), paroxetine (20-1280 μg/kg), sertraline (20-1280 μg/kg), citalopram (20-1280 μg/kg) and 8-OH-DPAT (1.25-80 μg/kg) were dissolved in 0.9% saline.

2.4 Statistical analysis

Data acquisition and analysis was accomplished using a PC Pentium and an integrated software package for electrophysiology (RISI, Symbolic Logic, Dallas, TX). Dose-response curves were constructed by comparing the mean firing rate during 2 min, starting immediately after the injection of each dose, with the basal firing rate. The data obtained with SSRIs were subjected to an analysis of variance (ANOVA) for repeated measures. Burst analysis of dopaminergic neurons was performed by using the RISI program running on a PC computer. A total of 500 consecutive spikes was recorded for each neuron before and at the peak of drug effect. Burst were defined in the standard fashion, starting with an ISI < 80 ms and ending with the concurrence of two spikes with an ISI > 160 ms [5]. Comparisons between groups were performed with Student's t test, while the increase of percentage of spikes occurring in bursts for neurons belonging to subclass A was tested through ANOVA analysis.

2.5 Processing analysis

The behavior of dopaminergic basal activity was investigated by building the function $\Delta t(n)$ (n: 1..500); where $\Delta t(n)$ indicates the temporal distance between the nth and $(n+1)^{th}$ spike; $\Delta t(n)$ is equivalent to the interspike interval function. Fig. 1A shows $\Delta t_j(n)$ functions for eight neurons. For each cell the basal firing pattern of 501 spikes recorded immediately before drug administration was used for the signal processing described below, 301 of the 501 spikes are shown on the left side of Fig. 1. The neurons have been organized in Fig. 1 to illustrate notable evidence of two different behaviors. Neurons showing an apparently more regular behavior of $\Delta t_j(n)$ functions are shown at the top of the figure, followed by neurons showing an apparently less regular behavior of the same functions. Less evident is the different degree of regularity among the different neurons. The following functional operator, related to the density-power-spectrum, expresses a quantitative measure of the disorder or entropy of the signal [8,13]:

$$\Psi(\Delta t_j(n)) = \sum_{k=0}^{N-1} \frac{|F(k)|}{F(0)} \tag{1}$$

where

$$F(k) = \sum_{n=1}^{N} \Delta t_j(n) \cos\left(\frac{2\pi nk}{N}\right) - i\Delta t_j(n) \sin\left(\frac{2\pi nk}{N}\right) \tag{2}$$

and consequently

$$F(0) = \sum_{n=1}^{N} \Delta t_j(n) \tag{3}$$

N = 500 [i.e. $\Delta t_j(N)$ indicates the temporal distance between the 500th and 501th spike equivalent to the last event considered]; the term **i** is the complex number $\sqrt{-1}$; the symbol $|\ |$ indicates the absolute value of the complex number F(k); F(k) is the Fast-Fourier-Transformation of $\Delta t_j(n)$ signal at the frequency k/N (or equivalently at the pulsation $2\pi k/N$); $|F(\bar{k})|$ represents the $\bar{k}^{th}$ component of the power-spectrum, that is a measurement of the energy at various frequencies; $|F(\bar{k})/F(0)|$ is the $\bar{k}^{th}$ component of the density-power-spectrum, i.e. power-spectrum normalized to the continuous or zero component of the signal.

Thus, Ψ in equation (1) represents the density-power-spectrum, defined as the sum of density-power-spectrum components, measured at each frequency. PC-Matlab User's Guide [11] has been used for computation of Ψ. The denominator F(0) in the

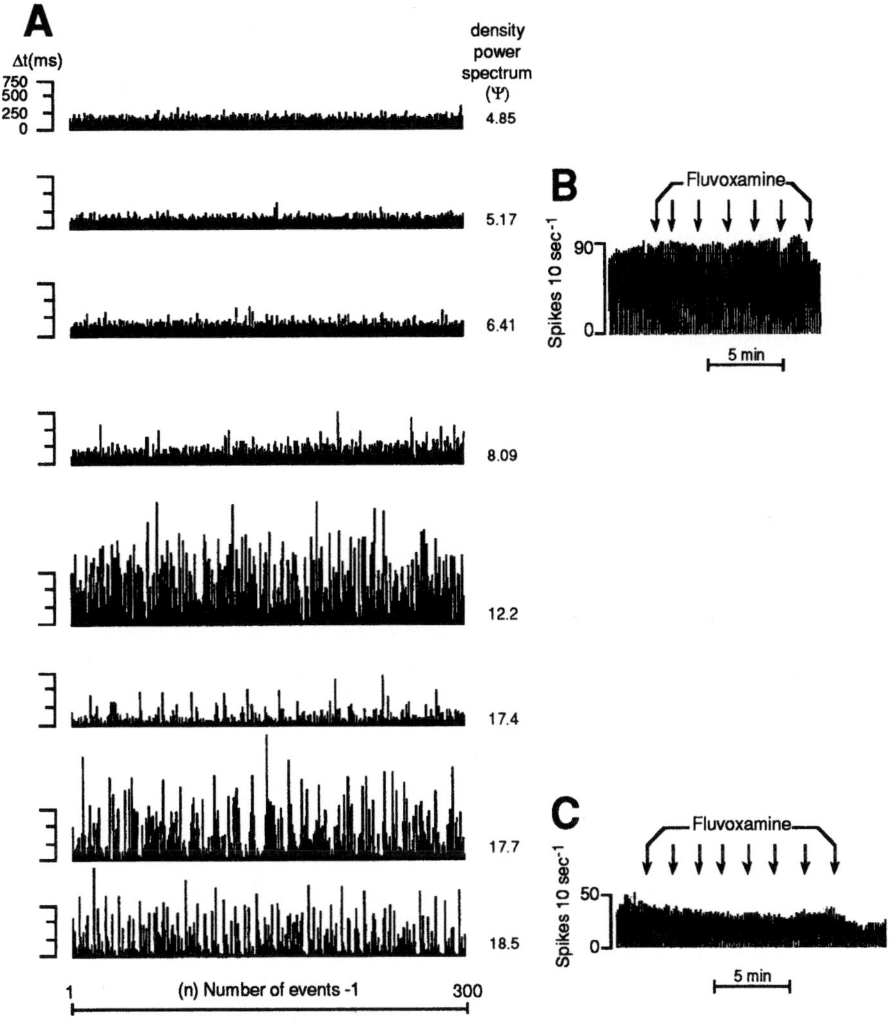

Fig. 1. Representative subsets of basal interspike interval progressions for eight different VTA DA neurons (A). The *x*-axis corresponds to interspike intervals. The sequence in which the cells have been listed follows the scale of power spectrum (Ψ). Representative rate histograms of two neurons with different Ψ values are displayed in B and C. The neuron shown in B belongs to subclass A: it shows a more regular behavior of interspike interval function $\Delta t_j(n)$ (the third row in A), had a lower value of Ψ function (6.41) and was not affected by fluvoxamine. The neuron shown in C belongs to subclass B: it showed a less regular behavior of ISI function $\Delta t_j(n)$ (the eighth row in A), had higher values of Ψ function (18.5) and exhibited a linear inhibition induced by fluvoxamine.

argument of series (1) is a factor of normalization. Thus, this operator Ψ considers as equivalent the behaviors of two firing activities with a constant envelope of $\Delta t_j(n)$ functions but different firing rates. The eight neurons in Fig. 1A have been displayed according to an increasing scale of Ψ function. The value of Ψ was computed for a temporal window representing the last 501 spikes before the first administration of selective 5-HT reuptake inhibitors and 8-OH-DPAT. The effect of fluvoxamine, paroxetine, sertraline, citalopram and 8-OH-DPAT was evaluated in terms of the percentage of variation of the basal firing rate. VTA DA neurons belonging to the experimental groups of fluvoxamine (n = 10), paroxetine (n = 12), sertraline (n = 10) and citalopram (n = 13), were subdivided into two subclasses: (A) neurons with no changes in their basal firing rate and (B) neurons showing an approximately linear inhibitory effect in response to selective serotonin reuptake inhibitors. VTA DA neurons belonging to the experimental group of 8-OH-DPAT (n = 18) were subdivided into two subclasses: (A) neurons showing a linear inhibitory effect or no changes in their basal firing rate and (B) neurons showing an evident excitatory effect (>10%) in response to 8-OH-DPAT.

3 Results

3.1 Effect of selective 5-HT reuptake inhibitors on the firing rate of VTA DA neurons

The neurons belonging to subclass A, whose firing rate was not affected by the drugs showed a more regular behavior of $\Delta t_j(n)$ functions and lower Ψ values (Fig. 1A, Ψ 4.85-8.09). The neurons belonging to subclass B, which exhibited an approximately linear inhibition by selective 5-HT reuptake inhibitors show a less regular behavior of $\Delta t_j(n)$ functions and higher Ψ values (Fig. 1A, Ψ 12.2-18.5). In Fig. 1B and 1C rate histograms are shown with a higher and lower Ψ value, respectively.

Fig. 2 shows a positive correlation between the level of disorder of basal activity expressed by the functional operator Ψ (lighting filling) and the corresponding reduction of the basal firing rate (dark filling) measured at the last cumulative dose of selective 5-HT reuptake inhibitors, equivalent to the maximal inhibitory effect for all the neurons selected (Spearman correlation: $r = 0.888$ with $P < 0.01$ for fluvoxamine; $r = 0.891$ with $P < 0.01$ for paroxetine; $r = 0.862$ with $P < 0.01$ for sertraline; $r = 0.885$ with $P < 0.01$ for citalopram). There was no significant difference between the mean basal firing rates of neurons belonging to subclass A and the neurons belonging to subclass B in the experimental groups of fluvoxamine (Student's *t*-test: $P > 0.05$), paroxetine (Student's *t*-test: $P > 0.05$) or citalopram (Student's *t*-test: $P > 0.05$). In contrast, in the sertraline's group, neurons belonging to subclass A showed a mean basal firing rate higher than those belonging to subclass B (Student's *t*-test: $P < 0.05$).

There was no significant difference (Student's *t*-test: $P > 0.05$) between the basal values of percentage of spikes occurring in bursts of neurons belonging to subclass A or subclass B. Overall, fluvoxamine, paroxetine, sertraline and citalopram did not cause any changes in the firing pattern, expressed in terms of percentage of spikes occurring in bursts of VTA dopaminergic neurons. However, in neurons belonging to subclass A, fluvoxamine, paroxetine, sertraline and citalopram induced a dose-dependent increase in percentage of spikes occurring in bursts (Table 1).

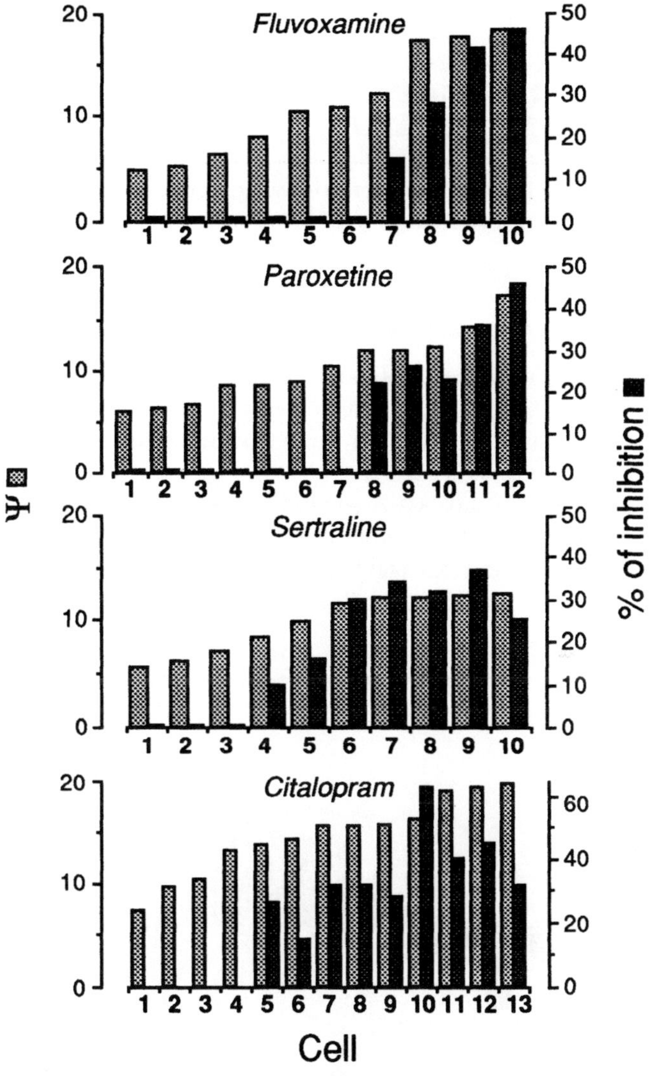

Fig. 2. Histograms showing the positive correlation existing between the level of disorder of basal activity of VTA DA neurons expressed by the power spectrum (Ψ) value and the corresponding reduction of the mean basal firing rate (expressed as % of inhibition) measured at the last cumulative dose of selective 5-HT reuptake inhibitors, equivalent to the maximal inhibitory effect for all the neurons selected. (Spearman correlation: $r = 0.888$ with $P < 0.01$ for fluvoxamine; $r = 0.891$ with $P < 0.01$ for paroxetine; $r = 0.862$ with $P < 0.01$ for sertraline; $r = 0.885$ with $P < 0.01$ for citalopram). Numbers on the x-axis indicate each DA neuron tested.

Table 1. Effects of selective serotonin reuptake inhibitors on the percentage of spikes occurring in bursts (PSB) of ventral tegmental area dopamine neurons belonging to subclass A.

	Baseline	320	640	1280
Fluvoxamine	10.2±5.4	20.7±10.6*	22.4±10.0*	27.1±12.7**
Paroxetine	7.3±4.6	21.4±8.7	22.8±9.3*	20.7±9.4**
Sertraline	8.6±7.1	14.0±9.8	18.3±12.9	20.5±16.4*
Citalopram	11.9±5.3	26.4±14.1	35.8 ±13.6**	35.8±13.1**

Data represent means ± S.E.M. Burst firing analysis was performed as described in the text. **P* < 0.05, ***P* < 0.01 as compared to baseline (ANOVA, followed by Tukey's test)

3.2 Effect of 8-OH-DPAT on the firing rate of VTA DA neurons

VTA DA neurons belonging to the experimental group of 8-OH-DPAT (n = 18) were subdivided into two subclasses: (A) neurons showing a linear inhibitory effect or no changes in their basal firing rate and (B) neurons showing an evident excitatory effect (>10%) in response to 8-OH-DPAT.

The neurons belonging to subclass A showed a less regular behavior of $\Delta t_j(n)$ functions and higher Ψ values. The neurons belonging to subclass B showed a more regular behavior of $\Delta t_j(n)$ functions and lower Ψ values. Fig. 3 shows a positive correlation between the level of disorder of basal activity expressed by the functional

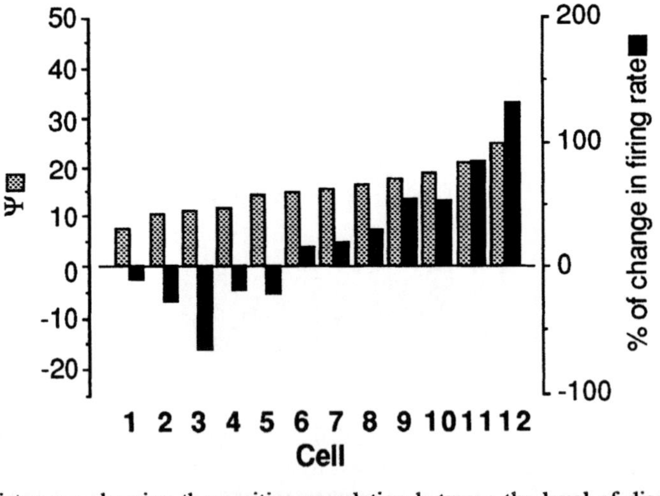

Fig. 3. Histogram showing the positive correlation between the level of disorder of basal activity expressed by the functional operator Ψ (lighting filling) and the corresponding change in mean basal firing rate [expressed as % of inhibition (negative) or excitation (positive)] (dark filling) (Spearman correlation: r = 0.83 with $P < 0.01$). Numbers on the *x*-axis indicate each DA neuron tested.

operator Ψ (lighting filling) and the corresponding change in mean basal firing rate [expressed as % of inhibition (negative) or excitation (positive)] (dark filling) (Spearman correlation: r = 0.83 with $P < 0.01$). There was no significant difference between the mean basal firing rates of neurons belonging to subclass A and the neurons belonging to subclass B (Student's *t*-test: $P > 0.05$). There was no significant difference (Student's *t*-test: $P > 0.05$) between the percentage of spikes occurring in bursts basal values of neurons belonging to subclass A or subclass B. A simple regression analysis on this set of data showed a higher correlation between basal firing properties of neurons and response to the drug when both power spectrum (Ψ) and firing rate (f) had been taken into account through their ratio. (linear correlation between % of change and Ψ: r = 0.68 with $P < 0.01$; linear correlation between % of change and (Ψ/f): r = 0.812 with $P < 0.01$).

4 Discussion

In this study a physical parameter, Ψ, was used to explain the differential response of VTA DA cells to selective inhibitors of 5-HT reuptake and to a 5-HT_{1A} receptor agonist. These data show that the density-power-spectrum represents a variable that can correlate the response of single DA neurons to agents modulating serotonergic transmission and, perhaps, to other classes of drugs. This parameter is also able to detect another difference in neuronal behavior. For example, subclass A neurons in the experimental groups of selective 5-HT reuptake inhibitors, characterized by lower values of Ψ showed a dose-dependent increase in the percentage of spikes occurring in bursts. This difference in electrical behaviors between subclass A and subclass B neurons could be related to different circuital properties.

It is well known that during *in vivo* electrophysiological experiments the response of single neurons to drugs or neurotransmitters may vary from no response to maximal inhibition or excitation. This variability might depend on the differential distribution of receptors on the somatodendritic area of the neuron sampled for recording. For example, it has been shown that the inhibitory response of VTA DA neurons to intravenously administered apomorphine and d-amphetamine as well as to microiontophoretically ejected DA depends on the basal firing rate of the neuron sampled [19]. There was a highly significant inverse relationship between basal activity and sensitivity to DA and DA agonists [19]. It was concluded that both spontaneous activity and sensitivity to DA agonists may be determined by the density (or sensitivity) of DA autoreceptors on VTA DA neurons [19]. As regards the effect of serotonergic drugs, it has been reported that electrical stimulation of the dorsal raphe selectively inhibited the firing rate of slowly firing (i.e. with a frequency below 4 spikes/sec) DA neurons in the substantia nigra, *pars compacta* [10]. Moreover, the 5-HT_{1A} agonist 8-OH-DPAT and 5-methoxy-N,N-dimethyltryptamine preferentially increased the firing rate of slowly firing DA neurons [10]. However, no correlation was found between the basal firing rate of VTA DA neurons and the excitatory effect of 8-OH-DPAT [16] or the inhibitory action of fluoxetine or citalopram [15].

The present study represents, to our knowledge, the first attempt to correlate the effects of serotonergic drugs on VTA DA neurons with their basal firing mode, as expressed by the new parameter Ψ. Unfortunately, it is presently impossible to establish the effect of chloral hydrate anesthesia on Ψ values, since the use of conscious paralyzed animals is not allowed in our laboratory. Nevertheless, it has been reported that chloral hydrate anesthesia exerts subtle effects on the basal activity and pharmacological responsiveness of midbrain DA neurons [9]. It is tempting to speculate that the basal firing mode of VTA DA neurons is dependent, among other factors, on the activity of serotonergic nerve terminals originating from the raphe nuclei [17]. It is known that about 50% of DA and non-DA neurons in the VTA receive synaptic contacts from serotonergic axon terminals [7]. It has been shown that 5-HT exerts an inhibitory influence on the activity of VTA DA neurons through the 5-HT_{2C} receptor subtype [16]. It is, therefore, possible that VTA DA neurons innervated directly or indirectly by the serotonergic system would have an irregular firing pattern and be inhibited by drugs enhancing synaptic levels of serotonin such as the selective 5-HT reuptake inhibitors or excited by drugs reducing synaptic levels of serotonin such as the selective 5- HT_{1A} receptor agonist 8-OH-DPAT .

Acknowledgments

This work has been partially supported by the Italian Research Council (Convenzione C.N.R. - Consorzio Mario Negri Sud).

References

1. Azmitia, E.C., Segal, M.: An autoradiographic analysis of the differential ascending projections of the dorsal and median raphe nuclei in the rat. J. Comp. Neurol. **179**, 641-668 (1978)
2. Bunney, B.S., et al.: Dopaminergic neurons: effect of antipsychotic drugs and amphetamine on single cell activity. J. Pharmacol. Exp. Ther. **185**, 560-571 (1973)
3. Grace, A.A., Bunney, B.S.: Nigral dopamine neurons: intracellular recording and identification with L-DOPA injection and histofluorescence. Science **210**, 654-656 (1980)
4. Grace, A.A., Bunney, B.S.: Intracellular and extracellular electrophysiology of nigral dopaminergic neurons. I: Identification and characterization. Neuroscience **10**, 301-315 (1983)
5. Grace, A.A., Bunney, B.S.: The control of firing pattern in nigral dopamine neurons: single spike firing. J. Neurosci. **4**, 2866-2876 (1984)
6. Grace, A.A., Bunney, B.S.: The control of firing pattern in nigral dopamine nerons: burst firing. J. Neurosci. **4**, 2877-2890 (1984)

7. Hervé, D., et al.: Serotonin axon terminals in the ventral tegmental area of the rat: fine structure and synaptic input to dopaminergic neurons. Brain Res. **435**, 71-83 (1987)
8. Kay, S.M., Marple, S.L.: Spectrum analysis, a modern perspective. Proc. IEEE **69**, 1380-1418 (1981)
9. Kelland, M.D., Freeman, A.S., Chiodo, L.A.: Chloral hydrate anesthesia alters the responsiveness of identified midbrain dopamine neurons to dopamine agonist administration. Synapse **3**, 30-37 (1989)
10. Kelland, M.D., Freeman, A.S., Chiodo, L.A.: Serotonergic afferent regulation of the basic physiology and pharmacological responsiveness of nigrostriatal dopamine neurons. J. Pharmacol. Exp. Ther. **253**, 803-811 (1990)
11. Moler, C., Little, J., Bangert, S.: PC-Matlab for MS-DOS Personal Computers-User's Guide. Sherbon: The Mathworks Inc. 1987
12. Mori, S., et al.: Light and electron microscopic immunohistochemical studies of serotonin nerve fibers in the substantia nigra of the rat, cat and monkey. Anat. Embryol. **176**, 13-18 (1987)
13. Oppenheim, A.V., Schafer, R.W.: Discrete-time signal processing. Engle Wood Cliffs: Prentice Hall 1989
14. Paxinos, G., Watson, C.: The Rat Brain in Stereotaxic Coordinates. New York: Academic Press 1986
15. Prisco, S., Esposito, E.: Differential effects of acute and chronic fluoxetine administration on spontaneous activity of dopaminergic neurons in the ventral tegmental area. Br. J. Pharmacol. **116**, 1923-1931 (1995)
16. Prisco, S., Pagannone, S., Esposito, E.: Serotonin-dopamine interaction in the rat ventral tegmental area: an electrophysiological study in vivo. J. Pharmacol. Exp. Ther. **271**, 83-90 (1994)
17. Steinbusch, H.W.M.: Serotonin-immunoreactive neurons and their projections in the CNS. Classical transmitters and transmitter receptors in the CNS. Part II. In: Björklund, A., Hökfelt, T., Kuhar, M.J. (eds.): Handbook of Chemical Neuroanatomy. Vol. 3. Amsterdam: Elsevier 1984, pp.68-125
18. Wang, R.Y.: Dopaminergic neurons in the rat ventral tegmental area. I. Identification and characterization. Brain Res. Rev. **3**, 123-140 (1981)
19. White, F.J., Wang, R.Y.: A10 dopamine neurons: role of autoreceptors in determining firing rate and sensitivity to dopamine agonists. Life Sci. **34**, 1161-1170 (1984)
20. Yeomanns, J. S., Maidment, N.T., Bunney, B.S.: Excitability properties of the medial forebrain bundle axons of A9 and A10 dopamine neurons. Brain Res. **450**, 86-93 (1988)

REGULATION BY INTRACELLULAR CALCIUM OF THE ACTIVITY OF $GABA_A$ RECEPTORS IN TWO DIFFERENT TYPES OF NEURONS

Aroldo Cupello [1], Holger Hydén[2], Maria V.Rapallino[1] and Mauro Robello[3]

[1] Centro di Studio per la Neurofisiologia Cerebrale, C.N.R., 16132 Genova, Italy

[2] Institute of Neurobiology, University of Goteborg, Sweden

[3] INFM, Dipartimento di Fisica, Università di Genova, 16100 Genova, Italy

Key words. $GABA_A$ receptors, Ca_i^{2+}, cerebellar granules, Deiters' neurons

1 Introduction

The levels of intracellular free calcium in nerve cells are critical for the activity of many enzymes and for several cell functions such as, for example, exocytosis [2] and its regulation [7]. Free calcium levels are regulated by buffering systems, compartmentation and extrusion from the cell. Calcium entry into nerve cells is also of importance in the regulation of gene expression. In particular in hippocampal cells it has been demonstrated that according to the route of calcium entry, different genes are activated [1].

An additional important role of intracellular Ca^{2+} in nerve cells is the regulation of neurotransmitter receptor activity. Thus, intracellular calcium accumulation via Ca^{2+} permeable non NMDA receptors inactivates neighbouring NMDA receptors [11,13]. NMDA receptors themselves appear to regulate their own function via calcium ions influx [19]. In addition, calcium ions contribute to the reduction of the NMDA component of excitatory autaptic currents of hippocampal neurons in culture, in the presence of micromolar concentrations of glutamate [27].

Referring to neuronal $GABA_A$ receptors, the general finding has been that increases in intracellular free calcium cause a decrease of $GABA_A$ receptor activity [9,20,25]. In addition, calcium ion influxes appear to mediate the inhibition of $GABA_A$ receptor activity caused by previous activation of NMDA receptors [3,15,17,23].

In the experiments reported here we studied the regulation of $GABA_A$ receptor function by intracellular Ca^{2+} in two different nerve cells types, cerebellar granule cells and the vestibular Deiters' neurons. This study was approached in the two cases with different techniques: the patch clamp electrophysiological recording technique in the first instance and a biochemical micromethod involving the determination of $^{36}Cl^-$ permeation across single membranes in the second one.

2 Materials and Methods

2.1 Cerebellar Granule Cells (Whole-cell patch clamp)

2.1.1 Cell Culture. Granule cells were prepared from 8 day old rats following the procedure of Levi et al.[12] as previously described [18]. Cerebellar granule cells were maintained in Basal Eagle's culture medium supplemented with 10% fetal calf serum, 25 mM KCl, 2 mM glutamine and 100μg/ml gentamicine. Cultures were treated with 10^{-5}M cytosine arabinoside from day 1 and experiments were performed in cultures between days 4 and 12 in vitro (4-12 DIV).

2.1.2 Solutions. The basic external solution consisted of (in mM): 137 NaCl, 5.4 KCl, 1.8 $CaCl_2$, 5 HEPES, 10 glucose. The pH was adjusted to 7.4 using NaOH. In some experiments 1 $MgCl_2$ was added or 1.8 $BaCl_2$ was substituted to 1.8 $CaCl_2$. The basic pipette filling solution contained (in mM): 142 KCl, 10 HEPES, 2 EGTA, 4 $MgCl_2$, 3 ATP Disodium Salt. The pH was adjusted to 7.3 using Trizma base. The following solutions were used when the effect of voltage-activated calcium channels was compared with the effect of NMDA. Extracellular (in mM): 132 tetraethylammonium chloride, 10 $CaCl_2$, 10 Glucose, 10 HEPES, 4-aminopyridine. The pH was adjusted to 7.4 using Trizma base. Intracellular (in mM): 142 CsCl, 4 $MgCl_2$, 10 HEPES, 2 EGTA, 3 ATP. The pH was adjusted to 7.3 using Trizma base. Glycine (10^{-5}M) was added to the external solution when NMDA was tested to facilitate expression of NMDA receptor mediated currents.
Extracellularly applied ligands and drugs [L-arginine, N_ω-nitro-L-arginine (NNA)], were diluted in bath solution and applied by a bath perfusion system (≈3 ml/min flow). Internally applied drugs adenosine-5′-O-3-thiophosphate (ATP-γ-S), deltamethrin 1R (DMSO solution 0.01%) and permethrin (DMSO solution 0.01%) were added to the pipette filling solution.

2.1.3 Chemicals. Deltamethrin 1R and permethrin were purchased from Alomonelabs, Jerusalem, Israel. Other drugs were from Sigma (St.Louis, MO).

2.1.4 Electrophysiology. Membrane currents were measured with the standard whole-cell patch-clamp technique by an EPC-7 (List-Electronic,Darmstadt, Germany). Patch electrodes were manufactured from borosilicate glass capillaries (Hilgenberg GmbH, Malsfeld, Germany) with a programmable Sachs and Flaming puller (model PC-84) with a final resistance of 4-10 MΩ when filled with the standard internal solution. Cell responses were filtered at 3 Khz and sampled at 100 Hz. Voltage stimulation and data acquisition were performed by a Labmaster D/A A/D converter driven by pCLAMP software (Axon Instruments, Burlingame, CA). Capacitance transient neutralization and series resistence compensation were optimized. Analysis was performed with pCLAMP and SIGMA PLOT (Jandel Scientific, Erkrath, Germany) software. Data are given as mean ± SD (cell number); the statistical significance of the difference between the data (*p*) was evaluated with the Student's t test.

2.1.5 Experimental procedure. The holding potential was set to -80 mV in all the experiments reported (unless otherwise specified) as this was the most suitable condition for recording the total chloride current. The cell capacitance was 3-5 pF and the input resistance was 1-1.5 GΩ. Run-down phenomena of chloride and calcium currents were prevented by the presence of ATP in the internal solutions. Chloride currents were elicited by 10^{-5} M GABA perfusion at two min intervals. After one min from GABA recovery, cells were perfused with 10^{-4} M NMDA for at least one min and then treated with 10^{-5} M GABA plus 10^{-4} M NMDA to avoid a rapid recovery from the NMDA effect. In some experiments we replaced the NMDA medium with the GABA solution (time needed about two seconds). In these cases the resulting inhibition of $GABA_A$ receptor function was the same (30±8% inhibition of $GABA_A$ response by NMDA in 4 cells compared to 32 ±9% in the presence of both GABA and NMDA in the same cells). Washing out NMDA ten seconds before GABA application resulted in only half the effect being found. Moreover, perfusion with glycine alone activated a small chloride current having no effect on I_{GABA} (n=4) and the application of NMDA alone was ineffective in 7 different cells (only in two cells was a 10% reduction was observed).

In some experiments, L-arginine or NNA were added to the culture medium for at least 30 min. Afterwards, cells were transferred to the basic external solution, containing the same drug, and patched. GABA receptor and NMDA receptor were activated and the NMDA effect on I_{GABA} calculated. In the case of L-arginine application, which reduced per se the GABA activated current, the NMDA effect was calculated by reference to the *residual* GABA current after L-arginine inhibition.

2.2 Deiters' neurons (Biochemical approach)

In these experiments we utilized male rabbits weighing 2.0-3.0 Kg. The preparation of single Deiters' membranes was performed as described previously [8,16]. The membranes were put on the gold plated glass with their extracellular side facing the upper microchamber M1 (Fig.1).

$^{36}Cl^-$ out→in permeation (M1→M2) was studied putting it in M1 at the radioactive concentration of 0.55μCi/ml and following its passage to M2 (intracellular side) in the microchamber system [8].

The incubations were performed for 30 sec at 30°C. After the incubation, aliquots were taken from M1 (10μl) and fron M2 (2μl) in that order for radioactivity counting.

For each animal routinely six membranes were prepared. Three of them were studied without GABA in M1 (extracellular side) and the average $^{36}Cl^-$ counts

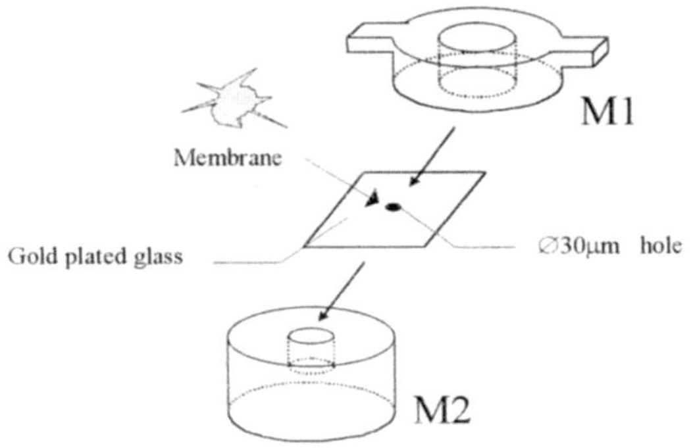

Figure 1. Microchambers scheme

passed to M2 (intracellular side) were taken as 1.0 (control). In the case of the other three membranes GABA was added in M1 and $^{36}Cl^-$ permeation was referred to that in the controls. Three such experiments were run for each GABA concentration in order to build concentration-response curves. Such curves were determined under different buffer conditions in M1 and M2 as detailed in Table 1.

Table 1

	M1 medium	M2 medium
1) First concentration response curve (Fig.5A)	NaCl 125 mM Na acetate 30mM K acetate 2.3mM Hepes 10mM; pH 7.2	NaCl 20 mM Na acetate 6.5mM K acetate 112.5mM Hepes10mM;pH7.2
2) Second concentration response curve (Fig.5B)	same as 1)	same as 1) plus: EGTA 1mM $CaCl_2$ 0.19mM
3) Third concentration response curve (Fig.5C)	same as 1) plus: $CaCl_2$ 1mM	same as 2)
4) Fourth concentration response curve (Fig.6A)	same as 1)	same as 1) plus: EGTA 1mM $CaCl_2$ 1.09mM
5) Fifth concentration response curve (Fig.6B)	same as 3)	same as 1) plus: EGTA 12mM
6) Sixth concentration response curve (Fig.6C)	same as 1) plus: $BaCl_2$ 1mM	same as 2)

3 Results

3.1 Granule Cells as Studied by Whole-Cell Patch-Clamp

3.1.1 Effect of Ca^{2+} Influx, Via either NMDA- or Voltage- Gated Channels , on $GABA_A$ Receptor Function

The currents activated by GABA on cultured granule cells under our basal perfusion and recording conditions, are mediated by $GABA_A$ receptors and involve Cl^- movements [18].

Previous activation of NMDA receptors by 100 μM NMDA and10μM glycine for 1 min (Fig.2A, second trace from the left) results in a reduced $GABA_A$ receptor effect while NMDA is still perfused (Fig. 2A, third trace vs. the first one). Removing NMDA+glycine the control GABA effect is regained after around 2 minutes (Fig. 2A, fourth trace).

The NMDA effect on GABA response was completely blocked substituting external Ca^{2+} with Ba^{2+} (6±1% inhibition vs. 40±% n=4). Similarly, addition of 1 mM $MgCl_2$ to the extracellular medium completely prevented the NMDA effect. Thus, the effect of NMDA receptor activation is evidently mediated by Ca^{2+} influx into the granule cells. This circumstance is confirmed by the fact that 10 mM BAPTA in the intracellular solution completely prevented the NMDA effect (0.5±0.3% inhibition, n=4).

The Ca^{2+} influx effect on $GABA_A$ receptors is specific for ions entering via NMDA channels, in fact even more than the double of Ca^{2+} charge entering via voltage activated Ca^{2+} channels does not result in any effect on $GABA_A$ receptor function (Fig.2B).

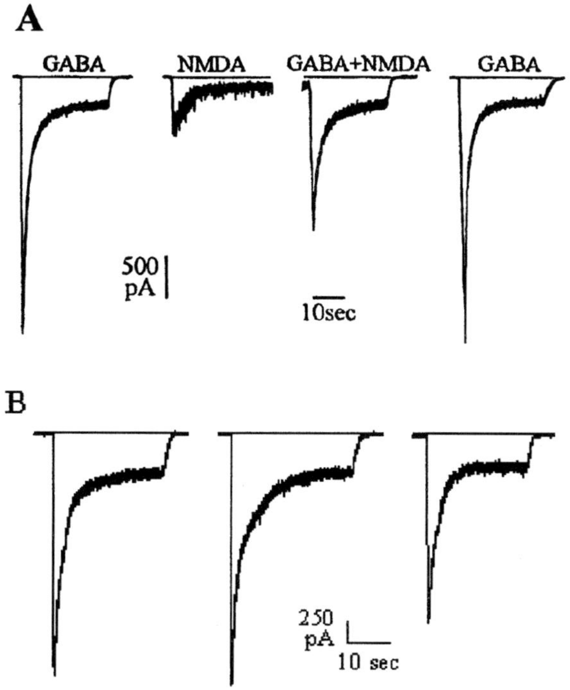

Figure 2. A) Effect of 10^{-4}M NMDA + 10^{-5}M glycine on GABA activated Cl^- current. The first trace represents the current activated by GABA before NMDA application, the second one represents NMDA current, the third one is GABA activated current in the presence of NMDA and the fourth trace represents the recovery. The traces were obtained at two minutes intervals. **B)** Lack of effect by 4 nCoulombs of Ca^{2+} entering via voltage activated channels. The first trace is the control GABA current, the second one was obtained after the Ca^{2+} current, the third one was again obtained after NMDA + glycine treatment. In all cases the holding potential was –80mV.

3.1.2 The NMDA Effect Involves, via Ca^{2+} Influx, Calcineurin and the Cascade NO-cGMP-PKG

In an effort to understand the mechanism(s) by which Ca^{2+} influx via NMDA channels influences $GABA_A$ receptors we tested whether a dephosphorylation mechanism via calcineurin could play any role [14]. Thus, we added ATP-γ-S, 3mM, in the pipette medium before applying NMDA/glycine. The idea was that

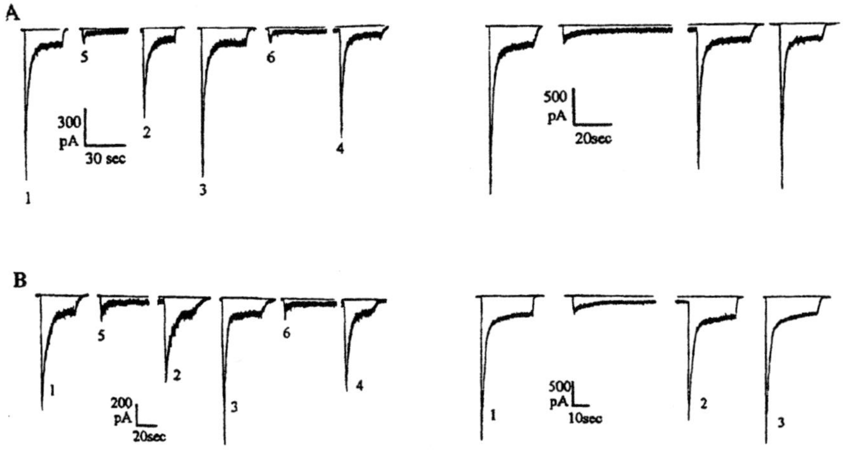

Figure 3. A Left) NMDA effect in the presence of 3 mM ATP-γ-S in the recording pipette. Traces 1, 5 and 2 represent respectively the GABA current, the NMDA current and GABA current in the presence of NMDA after 2 min of internal perfusion with ATP-γ-S. Traces 3, 6 and 4 represent the same currents after 30 min of internal perfusion with ATP-γ-S. **A Right)** Traces of $GABA_A$ response recorded before, during and after (wash-out) 10^{-4} M NMDA perfusion in a cell internally treated with $5x10^{-9}$ M deltamethrin. The second trace represents the NMDA activated current. **B Left)** NMDA effect in the presence of L-arginine. Traces 1, 5 and 2 represent respectively the control GABA current, the NMDA current and the GABA current in the presence of NMDA in a cell incubated with 10^{-4} M L-arginine for 30 min. Traces 3, 6 and 4 represent the same currents after wash-out of L-arginine. **B Right)** NMDA effect on 10^{-5} M GABA activated currents in a cell incubated with N_{ω}-nitro-L-arginine, $2x10^{-4}$ M, for 30 min. 1 is GABA activated current before NMDA treatment, 2 represents GABA current during NMDA treatment and 3 represents GABA current after NMDA wash-out.

since this ATP analogue provides non-hydrolysable phosphate groups on phosphorylated serine/threonine amino acid residues, this would prevent the NMDA effect if serine/threonine dephosphorylation was involved. This appears to be the case as shown by the results in Fig. 3A (left panel). In fact the presence of ATP-γ-S in the pipette medium erased around 50% of the NMDA effect under steady state conditions (around 20% inhibition vs. a control 40% inhibition). In addition, adding 5 nM deltamethrin, a calcineurin inhibitor [6], in the pipette medium again prevented half of the NMDA effect (Fig. 3A, right panel).

In other experiments we tested whether the L-arginine-NO-cGMP-PKG cascade was involved. Pretreatment of granule cells with 100μM L-arginine resulted in half of the effect by NMDA (20% inhibition of the GABA induced currents), as shown in Fig. 3B, left panel. Similarly, preincubation of the cells with the nitric oxide synthase (NOS) inhibitor, N_ω-nitro-L-arginine, 200 μM, prevented half of the NMDA effect (Fig. 3B, right panel).

3.1.3 The Calcineurin and NO-PKG Effects are Additive

In further experiments we tested whether by simultaneously blocking calcineurin and NOS one could prevent the NMDA effect. Two strategies were used to block calcineurin, an indirect one using ATP-γ-S in the recording pipette medium and a direct one adding to the pipette medium deltamethrin. In both cases, these treatments were coupled to the use of N_ω-nitro-L-arginine (200μM) in the perfusion medium.

Both deltamethrin plus NNA (Fig. 4A) and ATP-γ-S plus NNA (Fig. 4B) resulted in a complete block of the NMDA effect. This strongly indicates that the two Ca^{2+} activated pathways, calcineurin and NO-cGMP-PKG, act independently so that their effects are additive.

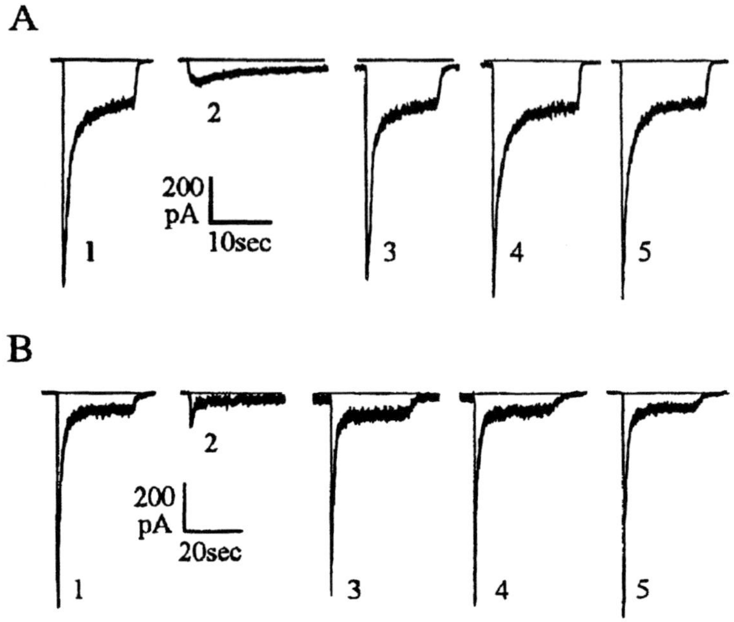

Figure 4. A) NMDA effect on GABA currents in the presence of NNA and deltamethrin. Traces represent: control (1), NMDA activated current (2), GABA activated currents after 1 min and 7 min of NMDA perfusion (3, 4) and after NMDA wash-out (5). **B)** NMDA effect on GABA currents in a cell incubated with $2x10^{-4}$ M NNA for 30 min and internally treated with 3 mM ATP-γ-S. Traces represent: control (1), NMDA ativated current (2), GABA activated currents after 1 min and 3 min of NMDA perfusion (3, 4) and after NMDA wash-out (5).

3.2 Deiters' Neuron Plasma Membrane and $^{36}Cl^-$ Permeation

3.2.1 Effect of " Intracellular " Ca^{2+} on GABA Stimulated $^{36}Cl^-$ Permeation

These studies were performed using the microchamber device shown in Fig.1. When we put $^{36}Cl^-$ in M1 it permeated to M2 across single Deiters' plasma membranes. "Extracellular" GABA (in M1) stimulated $^{36}Cl^-$ permeation with a peak (+ 37%) at [GABA] = 10^{-5}M when buffers nominally Ca^{2+} free were used (Fig. 5A). At higher GABA concentration the effect is lower probably because of receptors' desensitization.. However, when Ca^{2+} was present in the "intracellular" compartment at 0.02 μM, as controlled by EGTA, the peak was at [GABA] $=10^{-4}$M (Fig. 5B).

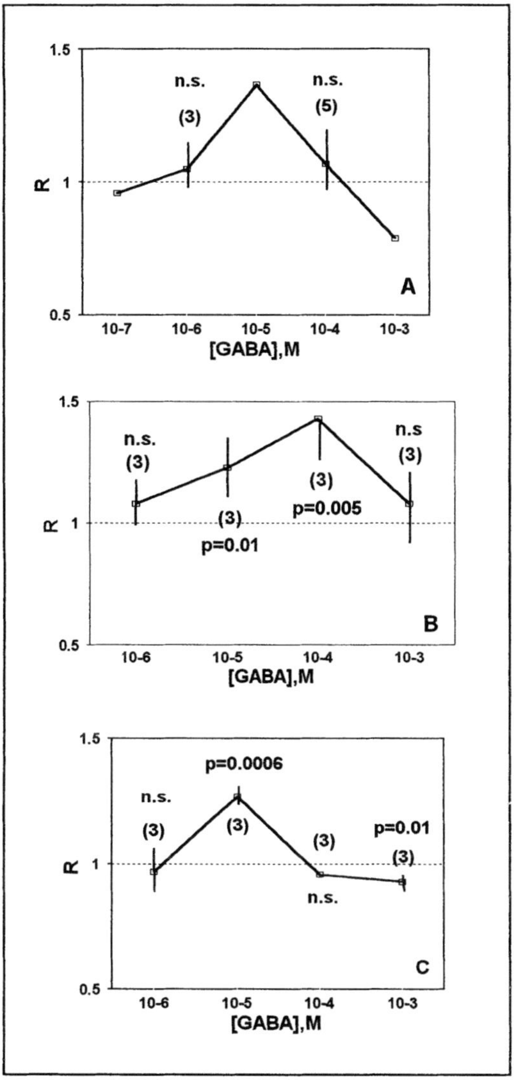

Figure 5:
A) Concentration-response curve for stimulation by extracellular GABA of $^{36}Cl^-$ out→in permeation. The permeation of $^{36}Cl^-$ in the presence of GABA at each concentration is referred to its passage in the absence of the neurotransmitter, taken as 1.0. The buffers in M1 ("extracellular" compartment) and M2 ("intracellular" compartment) were those reported in Table 1, line 1). **B)** Same as in case A) with the buffers reported in Table 1, line 2). **C)** Same as in case A). However, the buffers in microchambers M1 and M2 were those reported in Table 1, line 3). The numbers within parentheses represent the numbers of experiments: where they are not reported, two different experiments were performed and averaged; p values were evaluated by the Student's t test.

A likely explanation of the change of the concentration-response curve profile is that in nominally Ca^{2+}-free buffers, where no EGTA is present, the actual $[Ca^{2+}]_i$ is in the micromolar range, as we determined by an atomic absorption technique. This relatively high $[Ca^{2+}]_i$ would favour $GABA_A$ receptors' desensitization. In fact, when we buffered Ca^{2+} at 0.02μM in the "intracellular" (M2) compartment but added 1 mM Ca^{2+} in the "extracellular" (M1) medium, the curve resembled the one found with unbuffered media (Fig. 5C).

3.2.2 Effect of High and Very Low "Intracellular" Ca^{2+} on GABA Stimulated $^{36}Cl^-$ Permeation

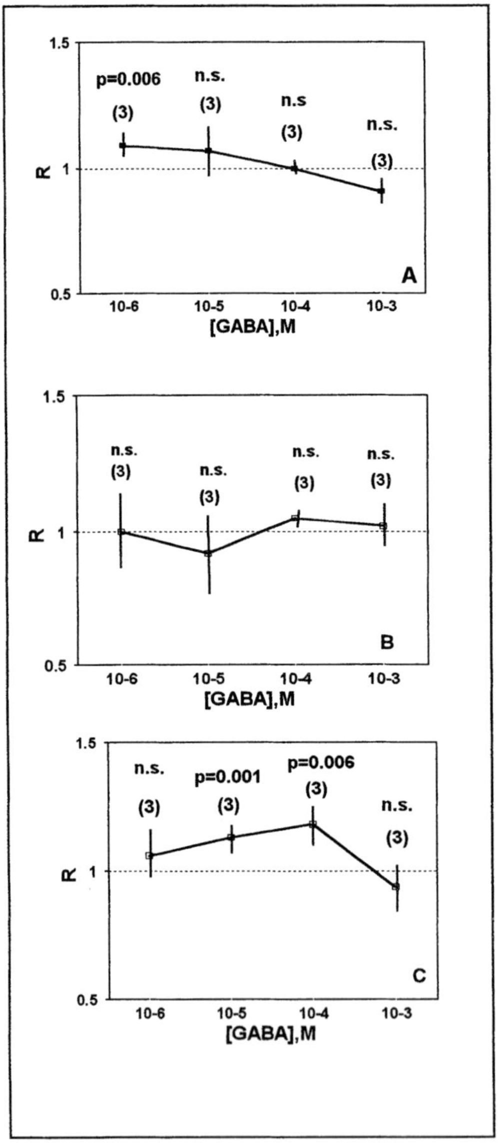

Figure 6:
A) Concentration-response curve for the stimulation by extracellular GABA of $^{36}Cl^-$ out→in permeation. The buffers in M1 and M2 were those reported in Table 1, line 4). **B)** Same as in case A). The buffers in M1 and M2 were those reported in Table 1, line 5). **C)** Same as in case A). The buffers in microchambers M1 and M2 were those reported in Table 1, line 6).
The symbols have the same meaning as in Fig. 5.

When we purposedly increased "intracellular" Ca^{2+} to the unphysiological level of 90 μM the GABA stimulation of $^{36}Cl^-$ permeation was completely erased (Fig. 6A). However, also the virtually complete removal of "intracellular" Ca^{2+} by the addition of 12 mM EGTA in M2 medium completely erased the GABA effect (Fig. 6B).

In a final experiment $[Ca^{2+}]_i$ was kept at the physiological level of 0.02 μM and 1 mM Ba^{2+} was added " extracellularly" in M1. In this case the stimulatory effect of

10^{-4}M GABA was maintained, although the overall curve was flatter than that obtained when neither Ca^{2+} nor Ba^{2+} were in the "extracellular" medium and $[Ca^{2+}]_i$ was 0.02μM (Fig. 6C versus Fig. 5B).

4 Discussion

4.1 Common vs. Specific Effects of $Ca^{2+}{}_i$ in the Two Neuronal Systems Studied

In both neuronal types studied, cerebellar granule cells of the rat and rabbit Deiters' neurons, increases in intracellular Ca^{2+} bring about a decrease in $GABA_A$ receptor function. The intracellular Ca^{2+} increase could not be quantified in the case of granule cells and resulted in a 40% reduction in $GABA_A$ whole-cell currents. In this case the route by which Ca^{2+} ions get into the cells is of decisive importance. In fact, even greater Ca^{2+} charge influxes via voltage activated Ca^{2+} channels do not end up with any change in $GABA_A$ receptor function. A possibility which we favour is that the NMDA receptor activity regulated $GABA_A$ receptors are situated in granule cells dendrites. Locally there the build up of $Ca^{2+}{}_i$ concentration via NMDA channel activity induces the cascades of events leading to a reduced $GABA_A$ receptor activity. Ca^{2+} influx via voltage gated channels involves mainly cell bodies and, in addition, may not be able to influence cell body $GABA_A$ receptors. In other words, the only 40% reduction in GABA activated whole-cell Cl^- currents we find is due to the fact that it involves only the fraction of $GABA_A$ receptors situated in the cells' dendrites, adjacent to NMDA glutamate receptors (Fig.7).

In the case of the Deiters' neurons, the experimental approach necessarily implies the study of cell body $GABA_A$ receptors. In this case the microchamber device allows the use of intra-cellular media of exactly known composition. The results show that comparatively high intracellular $[Ca^{2+}]$, in the low micromolar range, favours a more marked desensitization of $GABA_A$ receptors. Very high $[Ca^{2+}]_i$ (90μM) completely erases $GABA_A$ receptor activity.

It would appear then that whereas in Deiters' neurons cell body $GABA_A$ receptors are sensitive to increases in $[Ca^{2+}]_i$, the same does not apply to the corresponding cell body's receptors in cerebellar granule cells. As we pointed out above, this is not for granted since in the latter case we do not have an idea about the actual Ca^{2+} intracellular concentrations reached in the cell body under voltage dependent calcium channels activation.

In general, studies in various neuronal systems have shown that $[Ca^{2+}]_i$ increases bring about negative modulation of $GABA_A$ receptor function [9,20,25]. This is born out also by our data at least as far as dendritic $GABA_A$ receptors in cerebellar granule cells and cell body $GABA_A$ receptors in Deiters' neurones are concerned.

In the first case the mechanisms involved have been documented, namely Ca^{2+}_i mediated activation of calcineurin and of the cascade NO-cGMP-PKG [17]. The interaction of NMDA and $GABA_A$ receptors in granule cell dendrites is schematized in Fig. 7.

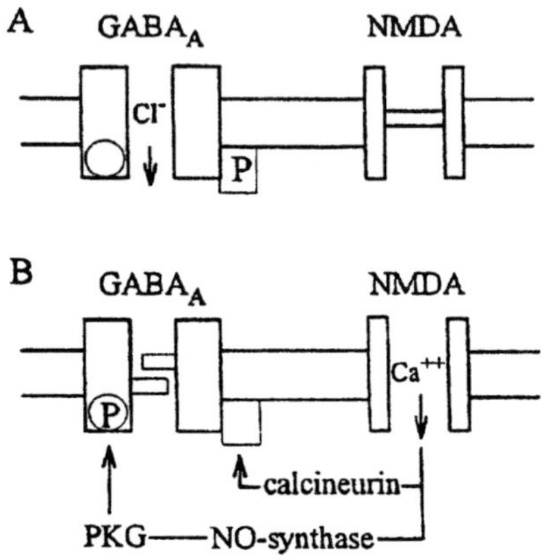

Figure 7: A possible schematic model of the NMDA mediated $GABA_A$ receptor modulation through a dual mechanism in cerebellar granule cells. The square closely apposed to the $GABA_A$ receptor represents a phosphorylated/dephosphorylated protein mediating its interaction with the cytoskeleton (see text). **A)** Under resting conditions binding of GABA to $GABA_A$ receptor results in channel opening. **B)** NMDA mediated calcium influx causes a partial inactivation of the $GABA_A$ receptor.

In the case of cell body $GABA_A$ receptor regulation by Ca^{2+}_i in Deiters' neurons the only conclusion about mechanism is that calcium favours such receptors' desensitization. However nothing can be said about how this is brought about. A surprising finding was that complete removal of Ca^{2+}_i by high intracellular EGTA in Deiters' neurons resulted in a complete block of $GABA_A$ receptors' activity (Fig. 6B). This does not happen in the case of granule cell $GABA_A$ receptors, judging from the fact that high (10 mM) BAPTA in the recording pipette not only did not impair $GABA_A$ receptors but prevented its block by NMDA receptor activation.

Referring to Deiters' neuron receptors, a conclusion which can be drawn from the data is that too a high or too a low $[Ca^{2+}]_i$ are harmful for a proper function of $GABA_A$ receptors. A similar finding has already been reported for such receptors in porcine pituitary intermediate lobe cells [25].

The different behavior of rat granule cell $GABA_A$ receptors could be due to a different subunit composition [24] leading to a receptor behaving in a different way from the point of view of regulation by Ca^{2+}_i.

Thus, the high $[EGTA]_i$ effect in the case of Deiters' neurons seems to imply that too low a free $[Ca^{2+}]_i$ is even more harmful to a proper $GABA_A$ receptor function in these neurons than low micromolar Ca^{2+}_i. However an alternative and paradoxycal explanation of such an effect is possible. In fact, it has been shown in rat vascular smooth muscle cells that an intracellular Ca^{2+} chelating agent increases Ca^{2+} influx, possibly due to removal of negative feedback by Ca^{2+} on its own entry [21]. It is thus possible that in the presence of very high $[EGTA]_i$ Ca^{2+} entry into the "intracellular" compartment (M2) from the "extracellular" one (M1), where it is at 1 mM, is more rapid. A Ca^{2+} rich layer just immediately below the neuronal membrane may form before being buffered by EGTA. This, in turn, may result in a complete erasing of $GABA_A$ receptor activity.

4.2 Physiological Implications

The NMDA receptor mediated impairment of $GABA_A$ receptors in granule cells underlines the existence of a direct cross-talk between excitatory and inhibitory events. A similar interaction, although samewhat different in mechanism, exist in hippocampal neurons [3,23].

NMDA receptors come into play when plastic modifications in neuronal behavior, such as long term potentiation [4], occur. On the other hand, hippocampal LTP induction is favoured by impaired inhibitory $GABA_A$ neurotransmission [26] and appears to be accompanied by a reduced inhibitory tone, at least in CA1 in the rat hippocampus [22]. Thus, one may speculate that impairment of $GABA_A$ receptors by NMDA activation is a built-in mechanism favouring the induction of plastic modifications. This could apply only to hippocampal neurons, in fact as far as we are aware no long term potentiation event has been so far documented for cerebellar granule cells.

In the case of the Deiters' neurons, the impairment of their $GABA_A$ receptors by high Ca^{2+}_i is in line with what has been generally found in other neuronal systems [9,20,25]. The only exception being cerebellar Purkinje cells where increases in Ca^{2+}_i bring about a stimulation of $GABA_A$ receptor activity [10]. This important exception may be related to the fact that these neurons undergo phenomena of long term depression, which is possibly accompanied by increased $GABA_A$ receptor function [5].

BIBLIOGRAPHY

1. Bading, H., Ginty, D.D., Greenberg, M.E. : Regulation of gene expression in hippocampal neurons by distinct calcium signaling pathways. Science 260, 161-165 (1993).

2. Burgoyne, R.D., Morgan, A.: Regulated exocytosis. Biochem. J. 293, 305-316 (1993).

3. Chen, Q.X., Wong, R.K.S.: Suppression of $GABA_A$ receptor responses, by NMDA application in hippocampal neurons acutely isolated from the adult guinea pig. J.Physiol. 482, 353-362 (1995).

4. Collingridge, G.L., Bliss, T.V.P.: Memories of NMDA receptors and LTP. TINS 18, 54-56 (1995).

5. De Schutter, E.: Cerebellar long -term depression might normalize excitation of Purkinje cells: a hypothesis. TINS, 18, 291-295 (1995).

6. Enan, E., Matsumura, F.: Specific inhibition of calcineurin by type II synthetic pyrethroid insecticides. Biochem. Pharm. 43, 1777-1784 (1992).

7. Greengard, P., Valtorta, F., Czernik, A.C., Benfenati, F.: Synaptic vesicles phosphoproteins and regulation of synaptic function. Science 259, 780-785 (1993).

8. Hydén, H., Cupello, A., Palm, A.: Gamma-amino-butyric acid stimulated chloride permeability across microdissected Deiters' neuronal membrane. Brain Res. 379, 167-170 (1986).

9. Inoue, M., Oomura, Y., Yakushiji, T., Akaike, N.: Intracellular calcium ions decrease the affinity of the GABA receptor. Nature 324, 156-158 (1985).

10. Kano, M., Rexhausen, U., Dreessen, J., Konnerth,A.: Synaptic excitation produces a long-lasting rebound potentiation of inhibitory synaptic signals in cerebellar Purkinje cells. Nature 356, 601-604 (1992).

11. Kyrozis, A., Goldstein, P.A., Heath. M.J.S., Mc Dermott, M.J.S.: Calcium entry through a subpopulation of AMPA receptors desensitized neighbouring NMDA receptors in rat dorsal horn neurons. J. Physiol. 485, 373-381 (1995).

12. Levi, G., Aloisi, F., Ciotti, M.T., Gallo V.: Autoradiographic localization and depolarization-induced release of amino acids in differentiating granule cell cultures. Brain Res. 290, 77-86. (1984).

13. Medina, I., Filippova, N., Barbin, G., Ben-Ari, Y., Bregestovski, P.: Kainate-induced inactivation of NMDA currents via an elevation of intracellular Ca^{2+} in hippocampal neurons. J. Neurophysiol. 72, 456-465 (1994).

14. Nairn, A.C., Shenolikar, S.: The role of protein phosphatases in synaptic transmission plasticity and neuronal development. Curr. Opin. Neurobiol. 2, 296-301 (1992).

15. Ragozzino, D., Eusebi, F.: Inhibition of GABA and glycine responses by glutamate in rat hippocampal neurons. Brain Res. 628, 115-120 (1993).

16. Rapallino, M.V., Cupello, A., Hydén, H.: Direct evidence for the presence of $GABA_A$ receptors on the cytoplasmic side of the Deiters' neuron plasma membrane. Brain Res. 462, 350-353 (1988).

17. Robello, M., Amico, C., Cupello, A.: A dual mechanism for impairment of $GABA_A$ receptor activity by NMDA receptor activation in rat cerebellum granule cells. Eur. Biophys. J. 25, 181-187 (1997).
18. Robello, M., Amico, C., Cupello, A.: Regulation mechanisms of GABA receptor function in cerebellar granule cells in culture: possible differential involvement of kinase activities. Neuroscience 53, 131-138 (1993).
19. Rosenmund, C., Feltz, A., Westbrook, G.L.: Calcium-dependent inactivation of synaptic NMDA receptors in hippocampal neurons. J. Neurophys. 73, 427-430 (1995).
20. Schwartz, R.D., Wagner. J.P., Yu, X., Martin, D.: Bidirectional modulation of GABA-gated chloride channels by divalent cations: inhibition by Ca^{2+} and enhancement by Mg^{2+}. J. Neurochem. 62, 916-922 (1994).
21. Skutella, M., Ruegg, U.T.: Increase of empty pool-activated Ca^{2+} influx using an intracellular Ca^{2+} chelating agent. Biochem. Biophys. Res. Comm. 218, 837-841 (1996).
22. Stelzer, A., Simon, G., Kovacs, G., Rai, R.: Synaptic disinhibition during maintenance of long-term potentiation in the CA1 hippocampal subfield. P.N.A.S. USA 91, 3058-3062 (1994).
23. Stelzer, A., Shi, H.: Impairment of $GABA_A$ receptor function by N-methyl-D-aspartate mediated calcium influx in isolated CA1 pyramidal cells. Neuroscience 62, 813-828 (1994).
24. Stephenson, F.A.: The $GABA_A$ receptors. Biochem. J. 310, 1-9 (1995).
25. Taleb, O., Trouslard, J., Demeneix, B.A., Feltz, P., Bossu, J.L., Feltz, A.: Spontaneous and GABA-evoked chloride channels on pituitary intermediate lobe cells and their internal Ca requirements. Pflugers. Archiv. 409, 620-623 (1987).
26. Wigstrom, A., Gustafsson, B.: Facilitation of hippocampal long-lasting potentiation by GABA antagonists. Acta Physiol. Scand. 125, 159-172 (1985).
27. Zorumski, C.F., Mennerick, S., Que, J.: Modulation of excitatory synaptic transmission by low concentrations of glutamate in cultured rat hippocampal neurons. J.Physiol. 494, 465-477 (1996).

Dynamics of Transmitter Release at CA3 Hippocampal Excitatory Synapses

Marco Canepari and Enrico Cherubini

Biophysics Sector and INFM unit, International School for Advanced Studies, Trieste, Italy

Abstract. Release of neurotransmitter is a stochastic process that can be studied by analysing whole cell postsynaptic currents evoked by repetitive activations of presynaptic fibres. In the case of the hippocampus, the stimulation of the Stratum Lucidum-Radiatum with a pipette filled with extracellular solution evokes excitatory postsynaptic currents (EPSCs) in CA3 pyramidal neurones. In the presence of $GABA_A$ receptors antagonists, in experiments in which 4 pulses at 50 ms intervals are delivered several times, synaptic responses show different behaviours (EPSC patterns). EPSC patterns can be classified in terms of potentiation or depression of the mean amplitudes of the second, third and fourth EPSC with respect to the first one. EPSC patterns in which the second and the third EPSCs are potentiated with respect to the first one are characteristic of synapses with a low release probability. In these synapses the residual calcium increases the release probability. In contrast, in synapses with a high initial release probability, vesicle depletion generates patterns in which all EPSC are depressed. A stochastic model that takes into account both changes in release probability and modifications in the number of synaptic vesicles available for exocytosis can account for the different synaptic behaviours that can be generated. This approach is a generalisation of the quantal analysis theory that can be used to investigate different features of synaptic function and plasticity.

Keywords. Hippocampus, Neurotransmitter release, synaptic activity, Stochastic process

1 Introduction

At fast synapses, the rise in intracellular calcium concentration, following an action potential, triggers a cascade of events leading to fusion, exocytosis of synaptic vesicles and release of transmitter into the synaptic cleft. This process occurs in a stochastic manner and the release of transmitter packets (quanta) fits a binomial distribution [14] [16]. When a second spike occurs in a nerve terminal within tens of ms after the first one, the synaptic conditions are changed by the first releasing event. Therefore, as a pair of action potentials occur in a nerve terminal, the mean current generated by the second spike is generally either potentiated or depressed with respect to the first one. The first phenomenon is called *pair-pulse facilitation* and the second one is called *pair-pulse depression*.

The phenomena of facilitation and depression, as well as the stochastic mechanisms of release, have been first characterised at the neuromuscular junction [3] [4] and later in CNS synapses. It has been suggested that facilitation is generated by an increase of release probability due to a higher presynaptic calcium concentration after the second spike (hypothesis of the residual calcium) [7]. In contrast, depression can be caused by depletion of the vesicles available for release after the first action potential [2]. Thus, synapses with a low probability of release are expected to be facilitated while synapses with a higher probability of release are more likely depressed as the number of available vesicles is generally reduced by the first releasing event.

When more than two spikes occur in presynaptic fibres, changes in release probability and in the number of available vesicle act together to determine the dynamics of synaptic responses and, in the case of long trains of action potentials, the mechanism of vesicle re-availability can also play a significant role. The dynamics of synaptic responses following a train of presynaptic action potentials has been recently studied in the neocortex by double recording from two connected neurones [9] [12]. In this system, it has been shown that excitatory synapses between pyramidal neurones are generally characterised by a depression of postsynaptic potentials (EPSPs) that follow the first one. The depression reaches a steady state value and this dynamic behaviour of synaptic responses can be modulated by synaptic plasticity [9].

In CA3 pyramidal neurones from transverse hippocampal slices, excitatory postsynaptic currents (EPSCs) generated by the repetitive activation of afferent fibres in the Stratum Lucidum-Radiatum show different dynamic behaviours. A stochastic model that takes into account the changes of release probability due to the residual calcium and the vesicle turnover has been developed in order to reproduce these behaviours. Changes of the extracellular calcium concentration and of the frequency of stimulation can modify the dynamics of synaptic responses as predicted by the model.

The analysis of synaptic responses generated by one or two action potentials has been used to demonstrate that the expression of many types of synaptic plasticity involves changes in the release probability [1] [8]. The analysis of synaptic responses generated by a train of presynaptic action potentials, as also proposed by others[10] [11], extends this approach to investigate also changes in the dynamic properties of synapses.

2 Experimental methods

2.1 Preparation and solutions

Synaptic responses were studied in transverse hippocampal slices (200 μm thick) cut from anaesthetised postnatal (P) days P7-P12 old Wistar rat brains. Slices were kept in a bathing solution containing (in mM): 126 NaCl, 3.5 KCl, 2 $CaCl_2$, 1.2 NaH_2PO_4, 1.3 $MgCl_2$, 14 $NaHCO_3$, 11 glucose, gassed with 95% O_2 and 5% CO_2 at 32 °C (pH 7.3). They were later transferred to the recording chamber in which they were superfused at 3ml/min. The extracellular solution contained (in mM): 126 NaCl, 3.5 KCl, 2 $CaCl_2$, 1.2 NaH_2PO_4, 1.3 $MgCl_2$, 25 $NaHCO_3$, 11 glucose, gassed with 95% O_2 and 5% CO_2 at room temperature (23 °C, pH 7.3). Bicuculline was also added routinely to the external solution in order to block $GABA_A$ mediated synaptic responses. To remove the NMDA component of the EPSCs, (+)-3-(carboxypiperazin-4-yl)-propyl-1-phosphonic acid (CPP) was added to the extracellular solution. In some experiments, extracellular calcium concentration $[Ca^{2+}]_o$ was then varied from 2mM to 4mM in order to increase the release probability.

2.2 Recordings

Recording pipettes were pulled from 2 mm borosilicate glass and had a resistance of 2-5 MΩ when filled with intracellular solution containing (in mM): 120 potassium gluconate, 4 $MgCl_2$, 49 N-2-hydroethylpiperazine-N'-2-ehanesulfonic acid (HEPES), 0.6 ethylene glycolbis (b-aminoethyl ether) N,N,N',N'-tetraacetic acid (EGTA), 2 Na_2ATP, 0.2 Na_2GTP, adjusted to pH 7.2 with KOH. Excitatory postsynaptic currents (EPSCs) from CA3 hippocampal neurones were recorded at -70 mV in voltage clamp mode, with a standard patch clamp amplifier (EPC-7 List Medical Instruments) after optimising capacitance and series resistance (typically 10-15 MΩ) compensation.

Patch pipettes with a tip of 10-30 μm diameter filled with the extracellular solution were used for stimulation. The stimulating electrode was placed under visual control on the surface of the slice in the Stratum Lucidum-Radiatum 50-200 μm above the CA3 pyramidal layer. EPSCs were evoked by four or ten pulses of currents (frequency 10-20 Hz, each pulse 40 μs duration), repeated every 10s. To stimulate the minimal number of fibres, the stimulus intensity was adjusted above threshold to evoke an EPSC. Typically the intensity of stimulation to evoke a minimal EPSC ranged from 4 to 10 V. Signals were low-pass filtered at 2 kHz and acquired at 10 kHz.

3 EPSCs evoked by four pulses of stimulation at 20 Hz

During a train of presynaptic action potentials, at each spike, the probability of release is changed by the variations of the active transient calcium that triggers release and the number of available vesicles is changed by the depletion and by the re-availability processes. The experimental situation in which EPSCs from CA3 pyramidal neurons are recorded when the Stratum Lucidum-Radiatum is stimulated by a train of 4 pulses (see figure 1.1 A), is described in figure 1.1 B.

As shown in the figure, the extracellular stimulation can generally produce the activation of one or more synaptic fibres converging onto the same postsynaptic neuron. In order to obtain the minimum number of stimulated fibres, the strength of the stimulating pulse must be the lowest to obtain evoked activity (minimal stimulation).

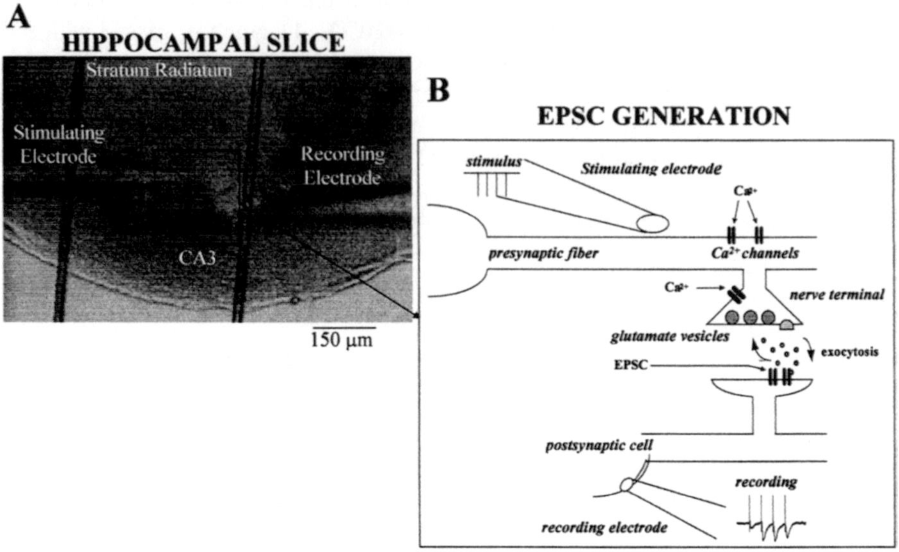

Figure 1.1. A. Hippocampal slice obtained from a P9 rat with the recording electrode positioned in the CA3 pyramidal region and the stimulating electrode placed in the stratum Lucidum-Radiatum. B. Illustration of the synaptic events underlying dynamics of synaptic responses evoked by a train of pulses.

The release is a stochastic process underlying the high variability of synaptic responses. The four traces of figure 1.2 show four single recordings of EPSCs evoked by four pulses of minimal stimulation at 20 Hz repetitively delivered every 10 seconds. The average of 40 recordings is also shown in the figure.
The variations in the mean EPSC amplitudes corresponding to the four pulses reflect changes of the release and possibly of the kinetics of postsynaptic receptors. In the case of two pulses of stimulation, these variations have been reported in literature as pair-pulse facilitation (if the mean of the EPSC amplitude evoked by the second stimulus is larger) and pair-pulse depression (if the mean of the EPSC amplitude evoked by the second stimulus is smaller).
The concepts of pair-pulse facilitation and pair-pulse depression can be generalised in the case of more than two presynaptic spikes. Different synaptic behaviours (EPSC patterns) can be classified and divided in families in terms of facilitation (F) or depression (D) of the nth mean EPSC with respect to the $(n-1)^{th}$ one. For example the synaptic behaviour illustrated in figure 1.2 in which a depression occurs after two facilitations belongs to the family of FFD EPSC patterns. In CA3 pyramidal neurons, different families of EPSC patterns have been observed.

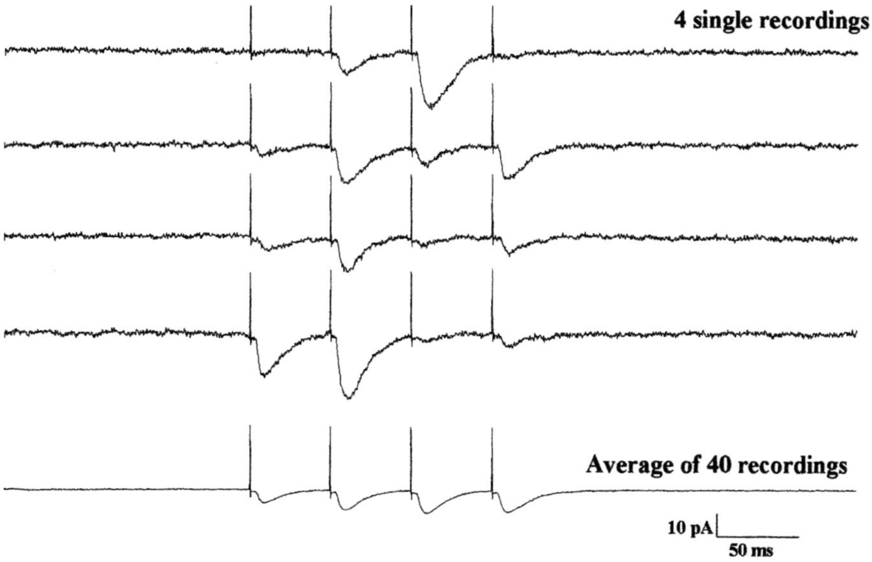

Figure 1.2. Four single recordings of EPSCs evoked by four pulses of stimulation at 20 Hz and the average of 40 recordings including the ones shown.

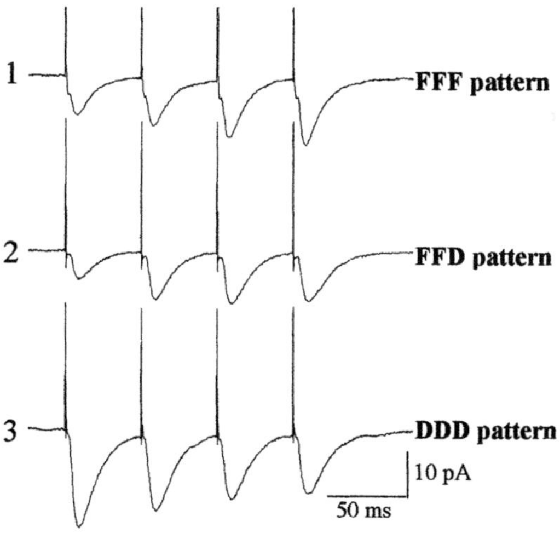

Figure 1.3. (1-3). Average of 40 recordings in three different experiments illustrating 3 cases of EPSC patterns belonging to the FFF FFD and FDD families respectively.

Figure 1.3 shows three examples of EPSC patterns belonging to the FFF, FFD and DDD families respectively. FDD and DDD patterns are the most common cases.

Different EPSC patterns can also be observed in the same cell by stimulating different areas of the Stratum Lucidum-Radiatum. When the initial probability of release is relatively low, the accumulation of residual calcium increases the release probability of the following spikes. On the other hand, when the release probability is already high from the beginning, the number of available vesicles is considerably reduced after the first spike. It follows that synapses with a low initial release probability are more likely facilitated while synapses with a high initial release probability are more likely depressed.

4 Theory of release dynamics

4.1 Transmitter release during a train of action potentials

Consider a synaptic terminal in which N vesicles are available for release and the process of release follows a simple binomial distribution with the same probability p for all the vesicles. Then the probability p(n) that n synaptic vesicles are released after an action potential is

$$p(n) = \frac{N!}{n! \cdot (N-n)!} \cdot p^{n} \cdot (1-p)^{N-n} . \qquad (1a)$$

It follows that the mean number of released vesicles $<n>$, its standard deviations σ_n and the probability to have a failure p(0) (release of 0 vesicles) are

$$<n> = p \cdot N \qquad (1b)$$

$$\sigma_n^2 = N \cdot p \cdot (1-p) \qquad (1c)$$

$$p(0) = (1-p)^{N} . \qquad (1d)$$

When more action potentials occur in a synaptic terminal within a short time after the first one, the conditions of release are generally changed. Particularly, the probability of release is different because of the effect of the previous calcium entry (residual calcium hypothesis), while the number of vesicles available for release is decreased by the number of vesicles that have been released after the first action potential and that have been not yet recycled.

The number of vesicles N_{j+1} available for the $(j+1)^{th}$ release is then given by N decreased by the numbers of vesicles $\tilde{n}_k$ (k=1,...,j) that have been released at the j previous action potentials and not yet recycled. Thus

$$N_j = N - \sum_{k=1}^{j-1} \tilde{n}_k \tag{2a}$$

The mean number of vesicles n_j released during the jth action potential is expressed by

$$<n_j> \; = \; <p_j \cdot N_j> = p_j \cdot (N - \sum_{k=1}^{j-1} <\tilde{n}_k>) \tag{2b}$$

It can be demonstrated that the variance $\sigma^2_{n_j}$ of the number of vesicles released during the jth action potential is expressed by

$$\sigma^2_{n_j} = <n_j^2> - <n_j>^2 = <p_j^2 \cdot N_j(N_j - 1) + p_j \cdot N_j> - p_j^2 \cdot <N_j>^2 \tag{2c}$$

or

$$\sigma^2_{n_j} = p_j \cdot (1 - p_j) <N_j> + p_j^2 \cdot (<N_j^2> - <N_j>^2) = \sigma_b^2 + p_j^2 \cdot \sigma^2_{N_j} \tag{2d}$$

where σ^2_b represents the variance of a binomial distribution of $<N_j>$ vesicles and $\sigma^2_{N_j}$ is the variance of N_j.

Finally, the mean probability $<p(0)>$ to have a failure during the jth action potential is expressed by

$$<p(0)> \; = \; <(1 - p_j)^{N_j}> \tag{2e}$$

In order to evaluate the possible dynamics generated by this stochastic process, it is necessary to explicit the time course of the release probability p_j and of the re-availability. A simple way to express the increase in the release probability due to the progressive accumulation of residual calcium, is to use the generalised classical Dodge-Rahamimoff equation [5] [15].

$$p_j = \frac{p_{max}}{1 + \{K \cdot ([Ca^{2+}]_{init} + \Delta_{j-1}[Ca^{2+}])\}^{-4}} \tag{3a}$$

where K is constant, p_{max} is the maximum value of p_j, $[Ca^{2+}]_{init}$ is the active intracellular calcium after one spike and $\Delta_{j-1}[Ca^{2+}]$ is the residual intracellular calcium after j-1 spikes. This probability function in which there is a summation of constant contributions to residual calcium ($\Delta_{j-1}[Ca^{2+}]=(j-1)\cdot\Delta[Ca^{2+}]$) can account for mechanisms of facilitation following a stimulation at a constant frequency.
A simple way to explicit the re-availability dynamics is to express $\tilde{n}_k$ as a fraction of n_k that continuously changes in time. Thus

$$\tilde{n}_k = f(t) \cdot n_k \qquad (3b)$$

where f(t) is time function (re-availability function) such that f(0)=1 and $f(t\to\infty)\to 0$. A first approximation of any f(t) is given by an exponential function with a time constant τ. By introducing this hypothesis, equation 2b becomes for a constant frequency of stimulation

$$< n_j > \; = p_j \cdot (N - \sum_{k=1}^{j-1} < n_k > \cdot f((j-k)\cdot\Delta t)) \,. \qquad (3c)$$

It is now possible to predict the EPSC pattern behaviour as a function of the release probability and of the frequency of stimulation by calculating $<n_j>$ from equation 3c. Figure 2.1 A shows the behaviour of EPSC patterns as a function of the release probability and of the residual contribution of calcium which is constantly kept as one third of the initial calcium.

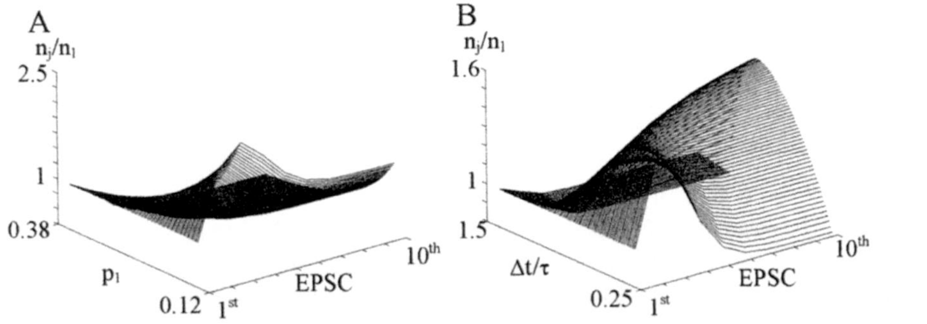

Figure 2.1. A. 100 EPSC patterns (ratios between n_j and n_1 with j=1,...,10) obtained by increasing the amount of initial and residual calcium in equation 3A. B. 100 EPSC patterns obtained by increasing the re-availability time (τ) and simultaneously decreasing the residual calcium in a synapse with an intermediate initial release probability.

This behaviour can be applied to the experimental situation in which the external calcium is increased or possibly to any situation of long term plasticity expressed by a presynaptic change of release probability. At low release probability, synaptic behaviours are dominated by facilitation of several currents with respect to the previous ones. As the release probability is increased, facilitations turn to depression until all the currents are depressed with respect to the previous ones.

Figure 2.1 B shows the behaviour of EPSC patterns as a function of the re-availability time constant and of the simultaneous decrease of the residual contribution of calcium. This behaviour can be applied to the experimental situation in which the frequency of stimulation is changed. As the frequency of stimulation decreases, the contributions to the residual calcium become negligible and an almost complete re-availability occurs between two consecutive releasing events. Thus the mean number of available vesicles becomes constant.

The evaluation of the variance and the mean probability to have a failure (equations 2d and 2e) can be achieved through a computer simulation of the stochastic process. It is useful to introduce the quantity

$$\rho_j = \frac{\sigma_j^2}{<N_j>} = (1 - p_j) + \rho_{N_j} \qquad (4)$$

In facilitatory synapses, the release probability increases from its low initial value. Therefore the first term of equation 4 can produce a slight decrease of ρ_j. On the other hand, if the initial probability is already high, the second term of equation 4 determines an increase of ρ_j. This behaviour of ρ_j is depicted in the simulations shown in figure 2.2 A

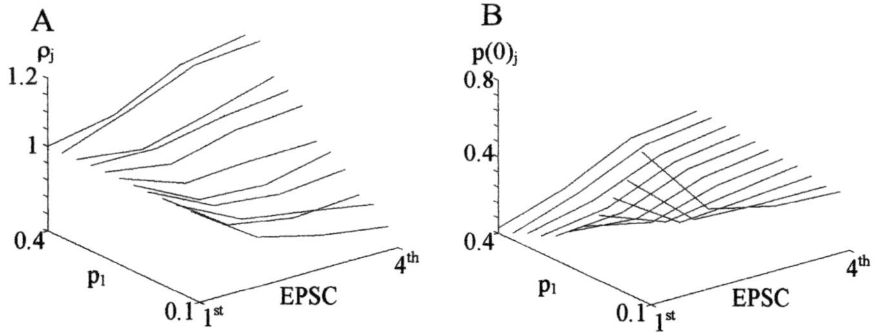

Figure 2.2. The results of computer simulations (10000 times) of the stochastic process of release with 5 vesicles obtained by increasing the amount of initial and residual calcium. A. The quantities $\rho_j=\sigma^2_j/<nj>$ (j=1,...,4). B. The probability to have a failure.

The probability to have a failure (shown in figure 2.2 B) strongly depends on the number of initial vesicles which is 5 in the case of the reported simulations and its time course is inverse to the time course of the mean amplitudes.

4.2 EPSC patterns at different external calcium concentrations and at different frequencies of stimulation.

A quantitative comparison between experimental data and the predictions of the theory can be obtained by recording evoked EPSCs in the presence of different calcium concentrations and at different frequencies of stimulation. The analysis of the time course of the variance and of the number of failures is not very significant for a low number of recordings (less than 50). However the EPSC patterns can be well reproduced by the model even with 30-40 recordings. Figure 3.1 A shows two averages of 40 recordings obtained in the presence of 2mM (A-1) and 4mM (A-2) calcium.

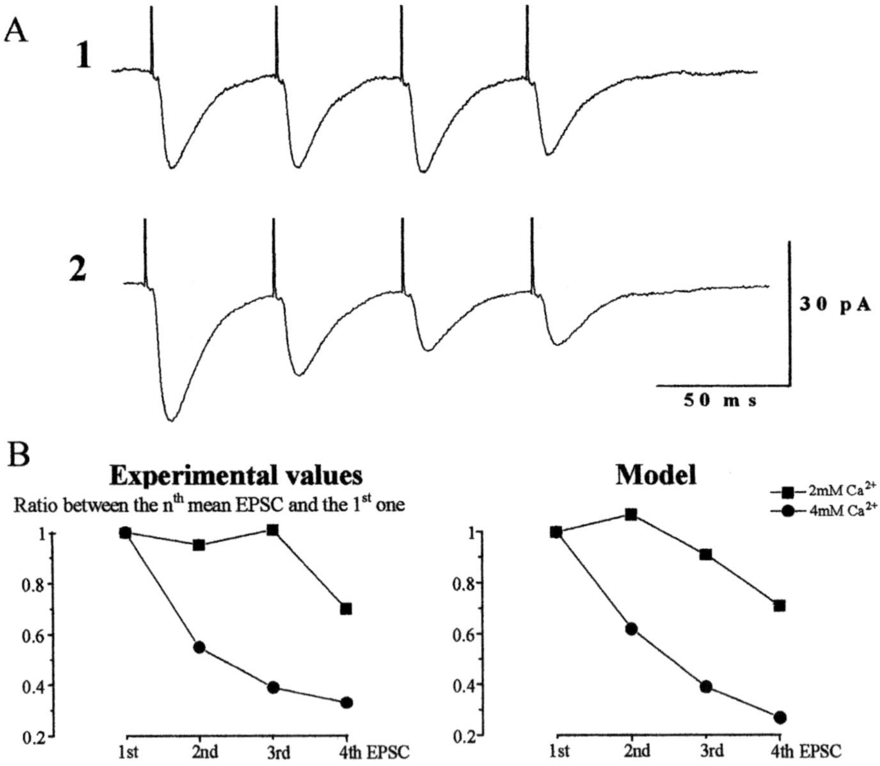

Figure 3.1. A. The average of 40 recordings in the same cell in the presence of 2 mM external calcium (1) and 4 mM external calcium (2). B. The comparison between the experimental EPSC patterns and the ones obtained with the model.

As the initial release probability is increased in high calcium concentration, the ratio between the nth EPSC amplitude and the first one decreases as predicted by the theory. Thus, in most of the cases, DDD EPSC patterns can be observed in the presence of 4mM calcium. In figure 3.1 B, the experimental EPSC patterns are compared with those obtained with equation 3c. The initial active intracellular calcium and its progressive increment have been increased in order to simulate the experimental condition of increasing the extracellular calcium.
In the experiment reported in figure 3.2, 10 pulses at 20 Hz (A-1) and 10 Hz (A-2) were used for stimulation. As predicted by the model, in the case of DDD EPSC patterns, the mean EPSC amplitudes are higher at lower frequencies of stimulation. This behaviour is shown in figure 3.2 B where the experimental EPSC patterns and those obtained by equation 3c are compared.

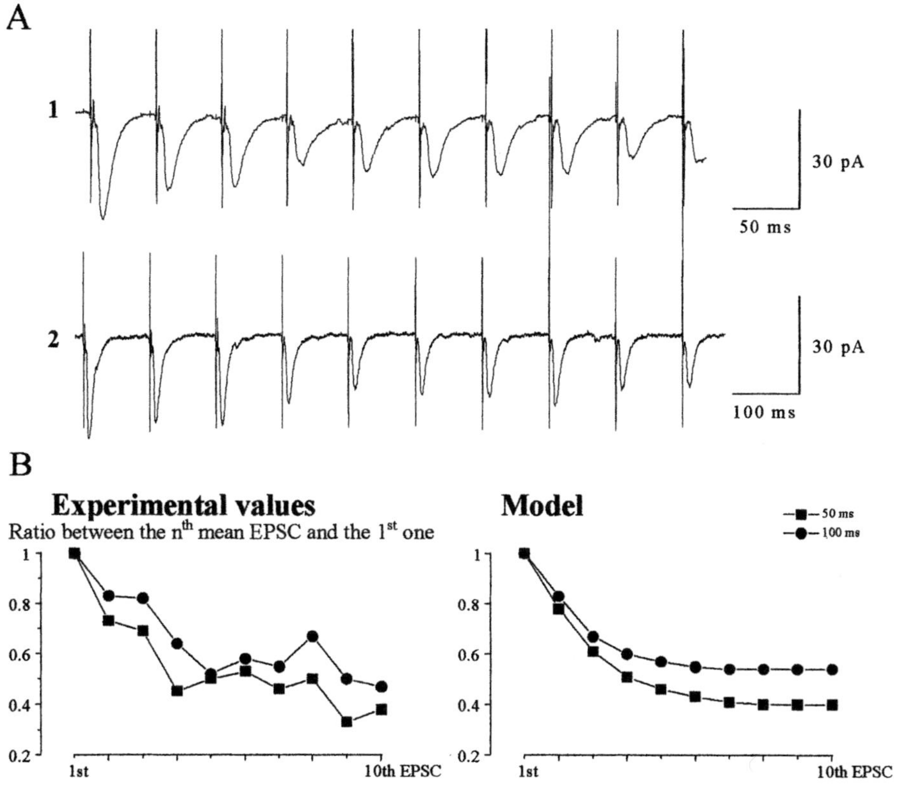

Figure 3.2. A. The average of 40 recordings with 10 pulses of stimulation at 50ms time interval (1) and 100 ms time interval (2). B. The comparison between the experimental EPSC patterns and the ones obtained with the model.

The re-availability time constant have been doubled and the progressive calcium increment have been decreased in order to simulate the experimental condition of increasing the interval between two pulses of stimulation.

5 Conclusions

Although the mathematical description and investigation on the main and physical principles of release are quite old topics of research in physiology, the study of the stochastic dynamics underlying synaptic activity elicited renewed interest among neuroscientists in the last years.

On one hand, the characterisation of the statistics of synaptic responses have been used to investigate changes of synaptic function. The analysis of postsynaptic response fluctuations evoked by stimulation of presynaptic fibres (Quantal Analysis) has been used to study the mechanisms of expression of long-term plasticity changes [6] [16]. From this point of view, the characterisation of the stochastic process of release induced by trains of presynaptic action potential generalises the traditional quantal analysis methods and offers a new tool to investigate also dynamic properties of synaptic function.

On the other hand, the process of release is the main synaptic mechanism underlying stochastic fluctuations of the activity of neural networks. These stochastic fluctuations can mask deterministic behaviours of nervous systems and can determine the space domain and the time domain of the neural code. From this point of view, the analysis of short term dynamics of synaptic activity is necessary to determine the properties of coding of synapse, as recently pointed out by Tsodyks and Markram [13], and is the first step to approach the more general problem of finding the spatial and temporal scales of neural coding in nervous systems.

In the hippocampus, the use of the analysis of the release dynamics to investigate the modifications of synaptic efficacy (long-term plasticity) and the activity of the network can be applied to the problem of understanding the functions of learning and memory processes.

Bibliography

1. Bekkers, J.M., Stevens, C.F.: Presynaptic mechanism for long-term potentiation in the hippocampus. Nature Lond. 346:724-729 (1990)

2. Bets, W.J.: Depression of transmitter release at the neuromuscular junction. J. Physiol. Lond. 206: 629-644 (1970)

3. Del castillo, J., Katz, B.: Quantal component of the end plate potential. J. Physiol Lond. 124: 560-573 (1954)

4. Del castillo, J., Katz, B.: Statistical factors involved in neuromuscular facilitation and depression. J. Physiol Lond. 124: 574-585 (1954)

5. Dodge, F.A., Rahamimoff, R.: Cooperative action of calcium ions in transmitter release. J. Physiol Lond. 193: 419-432 (1967)

6. Faber, D.S., Korn, H.: Applicability of the coefficient of variation method for analysing synaptic plasticity. Biophys. J. 60:1288-1294 (1991)
7. Katz, B., Miledi, R.: The role of calcium in neuromuscular facilitation J. Physiol Lond. 195: 481-492 (1968)
8. Manabe, T., et al.: A Modulation of synaptic transmission and Long-Term Potentiation: effects on paired pulse facilitation and EPSC variance in the CA1 region of the hippocampus. J. Neurophysiol. 70: 1451-1458 (1993)
9. Markram, H., Tsodyks, M.V.: Redistribution of synaptic efficacy between neocortical pyramidal neurons. Nature Lond. 382: 807-810 (1996)
10. Melkonian, D.S.: Transient analysis of a chemical synaptic transmission. Biol. Cybern. 68:341-350 (1993)
11. Quastel, D.M.J.: The binomial model in fluctuation analysis of quantal neurotransmitter release. Biophys. J. 72:728-753 (1997)
12. Thompson, A.M., West, D.C., Deuchars, J.: Properties of single axon excitatory postsynaptic potentials elicited in spiny interneurons by action potentials in pyramidal neurons in slices of rat neocortex. Neuroscience 63:727-738 (1995)
13. Tsodyks, M.V., Markram, H.: The neural code between neocortical pyramidal neurons depends on neurotransmitter release probability. Proc. Natl. Acad. Sci. *USA* 94: 719-723 (1997)
14. Van Der Kloot, W.: The regulation of quantal size. Progr. in Neurobiol. 36:93-130 (1991)
15. Van Der Kloot, W., Molgó, J.: Quantal Acetylcholine release at the vertebrate neuromuscular junction. Physiol. Rev. 74: 899-991 (1994)
16. Voronin, L. L.: Quantal analysis of hippocampal Long-Term Potentiation. Rev. in the Neurosci. 5: 141-170 (1994)

Responses of Isolated Olfactory Sensory Neurons to Odorants

Cristiana Picco[1], Paola Gavazzo[1], Stuart Firestein[2] and Anna Menini[1]

[1] Istituto di Cibernetica e Biofisica, CNR, Via De Marini 6, 16149 Genova, Italy
[2] Department of Biological Sciences, Columbia University, New York, NY 10027, USA

Keywords. Olfactory neurons, odorants, cyclic AMP

1 Introduction

The detection of odorant molecules begins in the nasal epithelium, where olfactory sensory neurons interact with odorant stimuli from the external world and produce electrical signals that are transmitted to the brain. Olfactory sensory neurons are small bipolar neurons with several fine cilia. The binding of odorant molecules to odorant receptor proteins in the cilia triggers an enzymatic cascade that leads to the onset of an inward ionic current, depolarization of the membrane and generation of action potentials that are conducted along the neuron's axon to the olfactory bulb, where further processing of odorant information occurs (for reviews, see Refs. [1,11,14,15]).

Odorant receptor proteins belong to the superfamily of receptors coupled to G-proteins. The activated receptor protein triggers a G-protein cascade and ultimately causes the activation of the effector protein adenylate cyclase that synthesizes cAMP. The increased cAMP concentration causes the opening of ion channels in the ciliary membrane [12], resulting in an odorant-induced current carried by sodium and calcium ions. Moreover calcium influx through the cAMP-gated channels triggers a calcium-activated chloride inward current [5,10]. Calcium ions play a very important role in olfactory signal transduction, affecting many elements of the transduction pathway and causing the olfactory sensory neuron to adapt [8,9,17].

In this chapter we examine the current response of isolated olfactory sensory neurons to various odorants and the blockage of the response by divalent cations and by odorants.

2 Materials and Methods

2.1 Olfactory sensory neuron preparation and current recording

Isolated olfactory sensory neurons were obtained from mechanical dissociation of the nasal olfactory epithelium of land phase adult salamanders *Ambystoma tigrinum*, previously decapitated and pithed. The normal Ringer solution contained (mM): NaCl, 115; KCl, 2.5; $MgCl_2$, 1.5; $CaCl_2$, 1; Hepes hemisodium, 9; pH 7.5. The 0 divalent solution had the same composition of normal Ringer except that the divalent salts $CaCl_2$ and $MgCl_2$ were not added. The intracellular pipette solution contained (mM): CsCl, 120; $MgCl_2$, 2; $CaCl_2$, 1; EGTA, 2; Hepes, 4; ATP, 1; GTP, 0.1; pH 7.5.

Currents were measured with an Axopatch 1D patch-clamp amplifier (Axon Instruments) in the whole-cell voltage-clamp mode at a holding potential of -55 mV, if not otherwise specified.

2.2 Odorants

The odorants geosmin, methyl geosmin, ethyl fenchol and fenchone were supplied by Givaudan-Roure Company. Cineole, acetophenone, guaiacol, citronellal and citralva were from Sigma and Aldrich. A 0.5 M stock solution of each odorant was prepared in dimethyl sulphoxide (DMSO) and diluted in the normal Ringer or in the 0 divalent Ringer solution to various concentrations.

2.3 Stimulus delivery

A rapid perfusion device was used for precise control of the ionic solution and of the odorant stimulus at the neuron surface. Four different solutions were flowing, side by side, into four parallel tubes at the back of the recording chamber. An isolated neuron, firmly attached at the tip of the patch pipette, was first placed in front of one pipe in a continuously flowing Ringer solution, and then rapidly moved in front of one of the adjacent pipes, where the test solution was flowing. Movement of the perfusion chamber was accomplished by a stepping motor under computer control allowing exposure times of variable duration to the various test solutions. The time required for changing solution was 30-50 ms, as estimated from the change in current of an olfactory neuron moved from Ringer to a solution containing an increased potassium concentration [2].

3 Results and Discussion

3.1 Odorant selectivity of individual olfactory neurons

The current in response to odorant stimuli was measured with the whole-cell patch-clamp technique in olfactory sensory neurons isolated from the salamander, while odorants at known concentrations were applied for controlled exposure times. It has sometimes been pointed out that the majority of odorants used for experiments with olfactory neurons of amphybians are not very likely to be present in the natural envinronment of these animals. We therefore also tested some odorants that are defined by humans as earthy odorants, like geosmin, methyl geosmin, ethyl fenchol, to check whether these odorants have a higher probability of being detected by salamander olfactory sensory neurons.

Mixtures composed of various odorants were applied to individual olfactory neurons. Experiments were done both in normal Ringer and in the absence of divalent cations in the Ringer solution to reveal any small odorant-induced current that could be obscured by divalent cation blockage (see Figs. 3.3 and 3.4). Olfactory neurons responding to an odorant mixture were subsequently tested with individual components of the mixture.

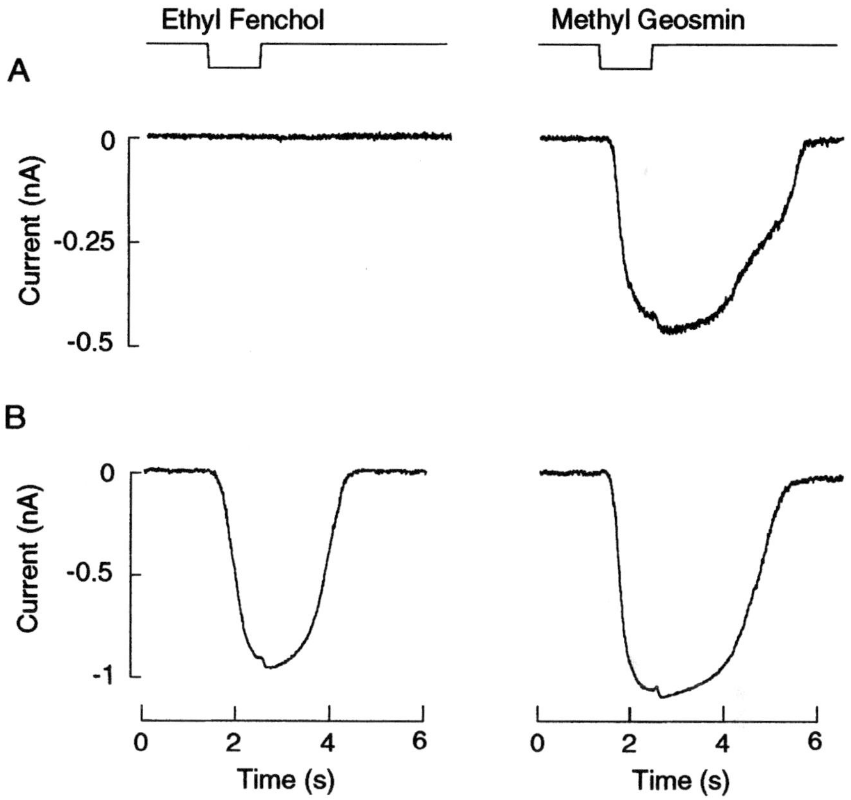

Fig. 3.1. Current responses of two individual olfactory sensory neurons to the earthy odorants ethyl fenchol and methyl geosmin. Odorants were applied at $5x10^{-6}$ M for 1.2 s. Holding potential was -55 mV. The Ringer solution did not contain divalent cations.

Fig. 3.1 shows current responses of two individual olfactory sensory neurons to the earthy odorants ethyl fenchol and methyl geosmin. The olfactory neuron in A responded only to methyl geosmin, while the neuron in B responded to both odorants. Forty neurons were tested for the response to a mixture of methyl geosmin and ethyl fenchol each present at a concentration up to $5x10^{-5}$ M and it was found that twenty-three neurons (58%) responded to the mixture with an inward current. Thirteen of the responding neurons were subsequently tested with individual components of the mixture and it was found that eleven neurons (85%)

responded to both odorants. Therefore the selectivity properties of olfactory sensory neurons for ethyl fenchol and methyl geosmin appear to be similar.

In a previous study [2] the mixture of acetophenone, amyl acetate and cineole was tested on several individual neurons and it was found that 54% of the neurons responded to this mixture. Moreover tests of the individual components on the olfactory neurons responding to the mixture, revealed that cineole was the preferred odorant, with 82% of the neurons responding at least to cineole. A comparison with the results obtained with earthy odorants shows that cineole, ethyl fenchol and methyl geosmin have a similar probability of eliciting a current response.

From electrophysiological experiments it is clear that an individual olfactory sensory neuron can detect many odorants [2,7,13]. However, studies of odorant receptor gene expression suggest that individual olfactory sensory neurons express one or at most a few odorant receptor genes (for reviews, see Refs. [1,14]). The molecular basis for the broad tuning of individual olfactory sensory neurons may be due to odorant receptor protein interacting with specific structural features (as for example functional groups, shape, charge distribution) of different odorants.

Several studies of the olfactory epithelium and bulb have provided a model for olfactory information processing (for reviews, see Refs. [1,11,14]), as illustrated in Fig. 3.2. Each olfactory sensory neuron expresses one odorant receptor which recognizes specifical structural features, or epitopes, in odorant molecules. Axons of neurons expressing the same receptor converge onto a single glomerulus. Therefore, for example, a given odorant A may have structural features that are recognized by odorant receptors OR1, OR3 and OR4 that activate the corresponding glomeruli. Another odorant B may have some different epitopes and activates for example odorant receptors OR1 and OR2. A third odorant may have completely different epitopes and activates OR5, OR6, OR7, OR8 and so on. Therefore an odorant receptor and its corresponding glomerulus can be activated by several different odorants, but each odorant is likely to elicit a unique pattern of glomeruli activation. This pattern may constitute an odorant-specific neural code.

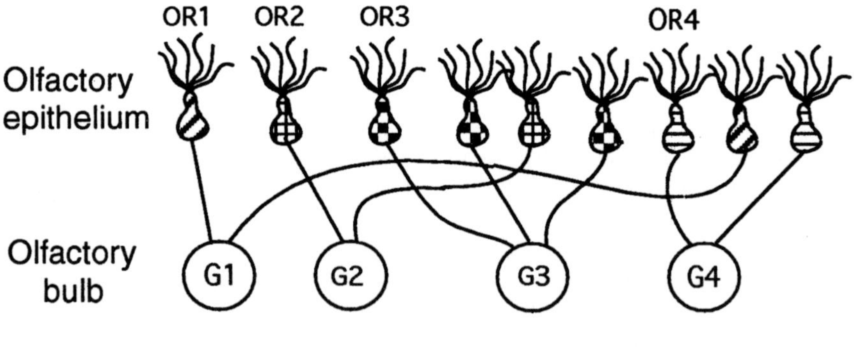

Fig. 3.2. Model for olfactory information processing. Olfactory sensory neurons expressing a given odorant receptor (OR) (filled with identical patterns) send their axons to a single glomerulus (G) in the olfactory bulb. Odorant molecules are characterized by their structural features (epitopes) and each type of odorant is recognized by the activation a unique combination of glomeruli.

3.2 Odorant response and current blockage

The odorant-induced current is composed of a cAMP-gated current carried by sodium and calcium ions and by a chloride current activated by the increased concentration of intracellular calcium entering the olfactory neuron through the cAMP-gated channels. cAMP-gated channels have been shown to be highly permeable to calcium ions, producing a large calcium influx during the odorant response [3]. Extracellular calcium ions not only permeate the cAMP-gated channels, but also partially block the current carried by sodium ions. Magnesium ions are also blockers of the cAMP-gated channels [3,4,16].

To investigate the effect of external divalent cations on the odorant response, an olfactory sensory neuron was first stimulated with an odorant-containing Ringer solution (Fig. 3.3A). At the end of the odorant stimulus, the olfactory neuron was immersed in a zero divalent Ringer solution (Fig. 3.3B). After removal of divalent cations the odorant-induced current suddenly increased almost two-fold, suggesting that many channels were activated by cAMP but that divalent cations partially blocked the inward current.

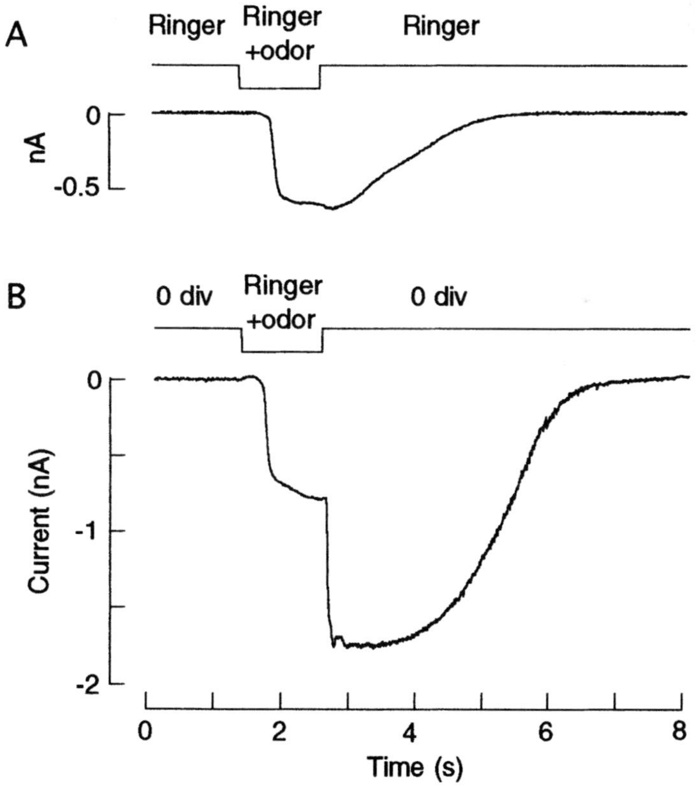

Fig. 3.3. Current responses of one olfactory sensory neuron in various ionic conditions to a mixture of the odorants methyl geosmin, ethyl fenchol and fenchone each present at 10^{-5} M. Holding potential was -55 mV. 0 div indicates that the Ringer solution did not contain divalent cations.

However, both odorant molecules and divalent cations were absent in the 0 divalent Ringer solution and it has been shown that some odorant molecules can partially block the current response [6]. For example at the end of the odorant stimulus in Fig. 3.3A the inward current had a very slight increase and then decayed back to baseline. The very small current increase is likely to be due to a partial unblockage of ion channels by the removal of the odorant molecules. Therefore from the experiments shown in Fig. 3.3 it is not possible to determine which percentage of the current unblockage was due to the removal of odorant molecules or of divalent cations, and it is very likely that both effects were present.

The blockage of the odorant-induced current by divalent cations and by odorants was further investigated.

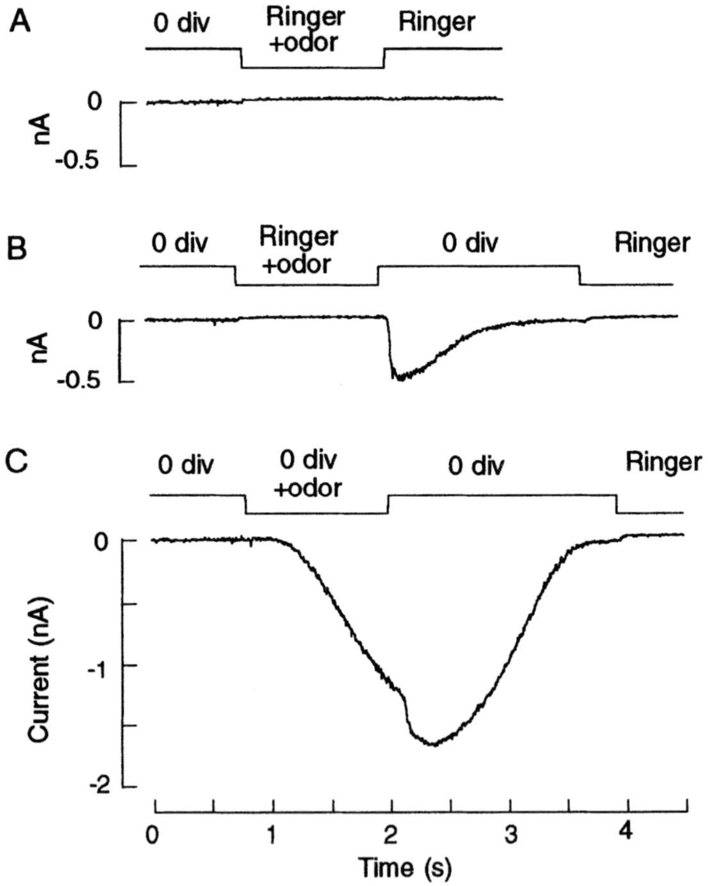

Fig. 3.4. Current responses of one olfactory sensory neuron in various ionic conditions to a mixture of the odorants citronellal, citralva and guaiacol each at a concentration of $5x10^{-5}$ M. Holding potential was -55 mV. 0 div indicates that the Ringer solution did not contain divalent cations.

Fig. 3.4 shows an olfactory neuron which did not respond with a measurable current to the odorant stimulus presented in Ringer solution (Fig. 3.4A). However, when normal Ringer was substituted with a 0 divalent Ringer solution, there was a sudden current increase (Fig. 3.4B). To distinguish between the odorant and the divalent cation blockage the same neuron was then maintained in a zero divalent Ringer solution (Fig. 3.4C) and the same odorant stimulus was applied in the absence of divalent cations. Fig. 3.4C shows that removal of divalent cations reveals a large odorant-induced current. This current was very likely entirely carried by Na ions through the cAMP-gated channels, since the absence of a Ca influx makes very unlikely the contribution to the current of Ca-activated Cl channels. At the end of the odorant stimulus, the inward current shows a further partial unblockage, due to the removal of the odorant molecules. These results indicate that in Fig. 3.4A the transduction pathway was activated by the odorant, but that both divalent cations and odorants completely blocked the measurable odorant-induced current.

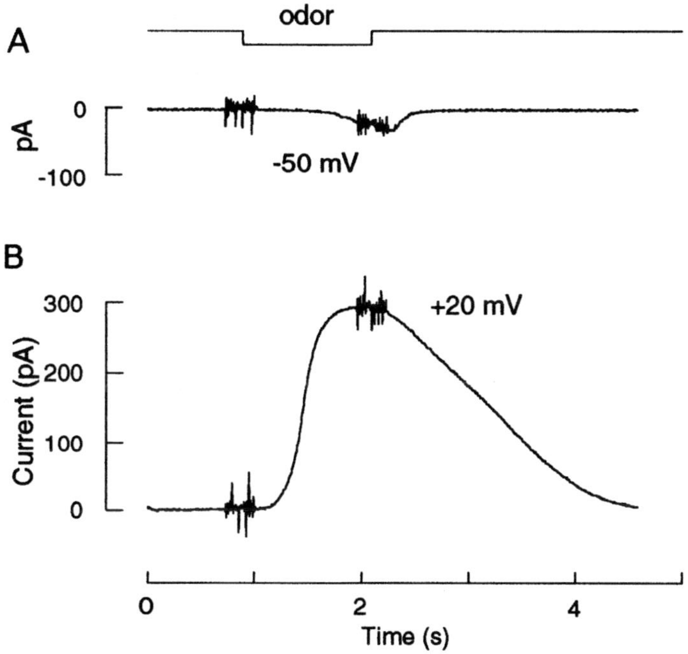

Fig. 3.5. Current responses of one olfactory sensory neuron to a mixture of the odorants cineole and acetophenone each at a concentration of $5x10^{-5}$ M. Normal Ringer solution was always present. Holding potential was -50 mV in A and +20 mV in B. An electrical artifact, caused by the stepping motor and indicative of the onset and offset of the stimulus, was present in each trace.

Another indication of the size of the divalent cation blockage is given by the experiments shown in Fig. 3.5. In these experiments the olfactory neuron was always kept in normal Ringer solution. A small odorant-induced current was measured at the holding potential of -50 mV (Fig. 3.5A). The holding potential was stepped at +20 mV and the same odorant stimulus as in A was applied (Fig.

3.5B) producing a large outward current. At positive membrane potential the electrochemical driving force for divalent cations is reduced and, since the divalent cation blockage of the cAMP-gated channels is voltage-dependent [3,4,16], a larger odorant-induced current was measured. At the normal resting potential, of about -55 mV, the channels are mostly blocked by physiological concentrations of calcium and magnesium ions, and therefore many channels will have to be activated to produce an inward current giving rise to depolarization.

Acknowledgments

We thank Avery Gilbert, from the Givaudan-Roure Company for supplying us with the odorants. We also thank G. Gaggero, P.G. Gagna, D. Magliozzi, and P. Guastavino for technical assistance. This work was carried out with support of EC grant BIO4-CT96-0593.

References

1. Buck, L.: Information coding in the vertebrate olfactory system. Ann. Rev. Neurosci. **19**, 517-544 (1996)
2. Firestein, S., Picco, C., Menini, A.: The relation between stimulus and response in isolated olfactory receptor neurons of the tiger salamander. J. Physiol. **468**, 1-10 (1993)
3. Frings, S., Seifert, R., Godde, M., Kaupp, U. B. : Profoundly different calcium permeation and blockage determine the specific function of distinct cyclic-nucleotide-gated channels. Neuron **15**, 169-179 (1995)
4. Kleene, S. J.: Block by external calcium and magnesium of the cyclic nucleotide-activated current in olfactory cilia. Neurosci. **66**, 1001-1008 (1995)
5. Kleene, S., Gesteland, R. C.: Calcium-activated chloride conductance in frog olfactory cilia. J. Neurosci. **11**, 3624-3629 (1991)
6. Kurahashi, T., Lowe, G., Gold, G. H.: Suppression of odorant responses by odorants in olfactory receptor cells. Science **265**, 118-120 (1994)
7. Kurahashi, T., Kaneko, A., Shibuya, T.: Ionic mechanisms of the olfactory transduction studied on isoltaed receptor cells of the newt. Neurosci. Res. Suppl. **12**, S85-S96 (1990)
8. Kurahashi, T., Menini, A.: Mechanism of odorant adaptation in the olfactory receptor neuron. Nature **385**, 725-729 (1997)
9. Kurahashi, T., Shibuya, T.: Ca-dependent adaptive properties in the solitary olfactory receptor cell of the newt. Brain Res. **515,** 261-268 (1990)
10. Kurahashi, T., Yau, K.-W.: Tale of an unusual chloride current. Curr. Biol. **4,** 256-258 (1994)
11. Moamberts, P.: Targeting olfaction. Curr. Opin. Neurobiol. **6**, 481- 486 (1996)
12. Nakamura, T., Gold G. H.: A cyclic nucleotide-gated conductance in olfactory receptor cilia. Nature **325**, 442-444 (1987)

13. Sicard, G., Holley, A.: Receptor cell responses to odorants: similarities and differences among odorants. Brain Res. **292**, 283-296 (1984)
14. Sullivan, S. L., Dryer, L.: Information processing in mammalian olfactory system. J. Neurobiol. **30**, 20-36 (1996)
15. Torre, V., Ashmore, J. F., Lamb, T. D., Menini, A.: Transduction and adaptation in sensory receptor cells. J. Neurosci. **15,** 7757-7768 (1995)
16. Zufall, F., Firestein, S.: Divalent cations block the cyclic nucleotide-gated channel of olfactory receptor neurons. J. Neurophysiol. **69**, 1758-1768 (1993)
17. Zufall, F., Shepherd, G. M., Firestein, S.: Inhibition of the olfactory cyclic nucleotide gated ion channel by intracellular calcium. Proc. R. Soc. Lond. B **246**, 225-230 (1991)

Part 2

Neuronal Networks

Long-Term Changes (LTC) of Evoked Field Potentials in the Amygdala: A Model of Emotional Memory.

Dan Yaniv and Gal Richter-Levin
Dept. of Psychology, University of Haifa, Haifa, 31905, Israel.

Abstract: The identification of the amygdala as an essential neural substrate for fear conditioning has permitted neurophysiological examinations of synaptic processes in the amygdala that may mediate fear conditioning. We examined synaptic plasticity in the thalamo-amygdala pathway, known to mediate acoustic fear conditioning and the hippocampal-amygdala pathway, involved in contextual fear conditioning. Results indicate that theta burst high frequency stimulation can induce either LTP or LTD in both pathways. The mechanisms responsible for long-term changes (LTC) in the amygdala are, to a great extent, elusive and further work is planned to elucidate them.

Keywords: Amygdala / long-term potentiation / long-term depression / emotional memory / fear conditioning / rat.

1. INTRODUCTION

It is well established that emotionally neutral stimuli can acquire the capacity to evoke striking emotional reaction following temporal pairing with an aversive event. Conditioning does not create new emotional responses but instead simply allows new stimuli to serve as triggers capable of activating existing, often hard-wired, species-specific emotional reactions. In the rat, for example, a pure tone previously paired with footshock evokes a conditioned fear reaction consisting of freezing behavior accompanied by a host of autonomic adjustments, including increases in arterial pressure and heart rate (5,24).

In addition to it's role in appetitive and perhaps attentional processes (e.g. ref. 14), converging evidence now indicates that the amygdala, and its many efferent projections, may represent a central fear system involved in the acquisition and expression of learned fear, both in experimental animals and humans (4, 11). The amygdala receives highly processed sensory information from all modalities through its lateral and basolateral nuclei (2, 25). In turn, these nuclei project to the central amygdaloid nucleus (1, 2, 32), which then projects to a variety of hypothalamic and brainstem target areas that directly mediate specific signs of fear and anxiety (cf ref. 11). Amygdala lesions, particular of the central nucleus, reduce or abolish various conditioned fear-related behaviors in a number of mammalian species. Acquisition and retention of learned fear responses such as an increase in blood pressure (16), changes in heart rate (19), fear-potentiated startle (15), and defensive freezing (16) are all attenuated by the lesion.

Selective lesions of structures afferent to the amygdala can also affect conditioning to specific sensory stimuli. For instance, lesioning of the medial geniculate nucleus (MGN) of the thalamus, which sends auditory information directly to the lateral (LA) and basolateral amygdaloid (BLA) nuclei, blocks the formation of tone-footshock, but not light-footshock, association as measured by conditioned freezing and blood pressure and the fear-potentiated startle response to an auditory stimulus (6, 25). Similarly, hippocampal lesions affect context-footshock association but not tone-footshock association (23, 30). The hippocampus is well known to play a key role in spatial memory and by way of CA1 and Subiculum areas also projects to the LA/BLA (31).

Numerous studies suggest that the N-methyl-D-aspartate (NMDA) class of excitatory amino acid receptors play an important role in fear conditioning. Administration of DL-2-amino-5-phosphonovaleric acid (APV), a selective NMDA antagonist, into the amygdala completely blocks the acquisition of fear-potentiated startle response to a light or tone stimulus (7, 29). A similar effect was observed when APV was microinfused into the BLA nucleus (a region rich in NMDA receptors), but not the central nucleus (a region sparse in NMDA receptors), of the amygdala using contextual fear conditioning and freezing (12). The drug APV, however, does not impair the performance of previously acquired fear responses such as freezing and startle (20, 29). Moreover, APV blocks fear conditioning only when administered prior to the time that the CS-US association occurs, but not immediately following the CS-US association (20). These findings suggest that NMDA receptors in the amygdala may be specifically involved in the fear learning processes. In addition, blockade of non-NMDA glutamate receptors (-amino-3-hydroxy-5-methyl-4-isoxazole-propionate or AMPA receptors) in the LA prevents the expression of learned fear (17, 21).

The dependence of fear learning on NMDA receptors in the BLA nucleus raises the possibility that a process like long-term potentiation (LTP), an enduring increase in synaptic transmission induced by high-frequency stimulation (HFS) of excitatory afferents, may be involved. Since APV affects amygdala-mediated fear conditioning and hippocampal LTP in similar manners (ie blocks acquisition/induction but not performance/expression) LTP has been proposed as a possible synaptic mechanism that mediates fear learning (20, 29). Clugnet and LeDoux (10) already demonstrated an LTP-like phenomenon in the auditory CS pathway from the thalamic medial division of the MGN to the LA. This auditory projection to the amygdala is also known to be glutamatergic (13). LTP has also been demonstrated in the BLA both in vivo (27), and using a brain slice preparation (9), although Chapman and Bellavance (8) have questioned the extant to which such plasticity in the amygdala can be selectively blocked by APV.

2. SYNAPTIC PLASTICITY OF THE THALAMO-AMYGDALA AND HIPPOCAMPAL-AMYGDALA PATHWAYS

The aim of the presented research was to characterize and compare in vivo activity- dependent synaptic plasticity in the thalamic MGN- BLA, and hippocampal subiculum-BLA pathways as a step in further elucidating memory-related synaptic plasticity in these pathways. By relating the amygdala, as a brain region closely related to a specific behavior (fear conditioning), with the possibility of inducing synaptic plasticity in that region, it is hoped to contribute to the investigation of synaptic plasticity in the formation of emotional memory.

Young male rats weighing between 250-350 g were anesthetized and prepared for recording. A recording glass electrode was stereotaxically positioned in the BLA and a bipolar stimulating electrode was inserted either into the MGN (n=10) or the subiculum (n=10). Low-frequency test pulses were delivered at 2 intensities: 20-40% of the maximum field potential size and 50-70% of it. LTP was induced by theta rhythm stimulation. 40 minutes post tetanization, responses to low-frequency test pulses were recorded, analyzed and compared to pre-HFS responses.

Results indicate the existence of synchronous cell activity at both pathways as field potentials were consistently achieved. Long-term synaptic changes (LTC - a change of more than 20 %) were evident in both routes although, surprisingly, in some cases, long-term depression (LTD) was induced by the same stimulating protocol which in other cases induced LTP (figure 1). Out of the 10 MGN-BLA animals, LTP was induced in 6 animals (95+/-30 %), LTD was evident in 2 (56+/-14 %) and 1 showed no LTC. The subiculum-BLA pathway seems less adequate for synaptic plasticity as only 5 out of 10 animals showed LTP (93+/-41 %), one showed LTD (43 %) and 4 animals showed no LTC.

3. DISCUSSION

The results obtained thus far indicate that both afferent tracts are amenable to LTC. Histological examination (not shown) did not provide any insight into how either LTP or LTD may be induced by the same stimulation protocol.

While considering possible explanations for this outcome, it is important to avoid attractive but inaccurate comparisons to hippocampal field potential research; the hippocampus is anatomically well-organized structure and the interpretation of the components of its field potentials are considered to directly reflect synaptic efficacy (EPSP slope) and cell excitability (population spike size) (3, 26). In contrast, the amygdaloid complex comprises of a heterogeneous group of cells without any particular internal anatomical organization. Thus, LTP of the field potential in the BLA can result from strengthening of excitatory synapses, depressing of inhibitory synapses, or both. The same idea appeals for BLA-LTD. Investigating the interrelations between principle cells and interneurons in the amygdala may contribute to a better understanding of both its activity and synaptic plasticity. In the hippocampus high-frequency stimulation results in LTP of principle (grannule) cells but also of interneuron synapses. This potentiation of inhibitory synapses alters local circuit activity (18, 28). We are now characterizing local circuit activity in the amygdala using stimulation protocols known to differentially activate hippocampal interneurons in order to further elucidate the mechanisms of synaptic plasticity in the amygdala.

4. SUMMARY AND CONCLUSIONS

Numerous studies in both rats and humans indicate the importance of the amygdala in the acquisition and expression of learned fear. The identification of the amygdala as an essential neural substrate for fear conditioning has permitted neurophysiological examinations of synaptic processes in the amygdala that may mediate fear conditioning. One candidate cellular mechanism for fear conditioning is long-term potentiation (LTP), an enduring increase in synaptic transmission induced by high-frequency stimulation of excitatory afferents. At present, the mechanisms underlying the induction and expression of amygdaloid LTP are only beginning to be understood, and probably involve NMDA and AMPA subclasses of glutamate receptors. In effort to enhance our understanding of the relationships of these processes to emotional (aversive) learning and memory, we examined synaptic plasticity in MGN-BLA pathway, known to mediate acoustic fear conditioning and Subiculum-BLA pathway, involved in contextual fear conditioning. Results indicate that high frequency stimulation can induce LTP and LTD in both pathways. However, the MGN-BLA tract seems to be more amenable to long-term changes (LTC) of synaptic transmission. Interpreting this data requires caution as the morphology of the amygdaloid complex in considerably less clear than that of the hippocampus. Thus, the mechanisms responsible for LTC in the amygdala are, to a great extent, elusive. Local-circuit activity examination, modified by pharmacological means, may be a powerful tool for the elucidation of these mechanisms, that are considered to be the biological core of emotional learning and memory.

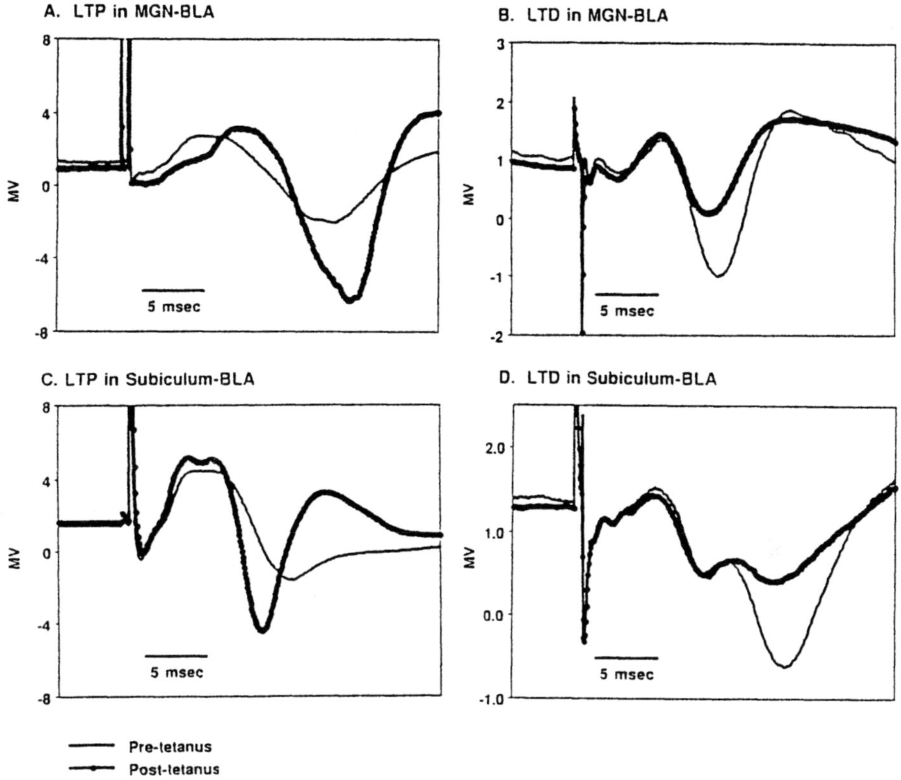

Figure 1: LTC in basolateral amygdala. LTP (A) or LTD (B) of the thalamo-BLA pathway and LTP (C) or LTD (D) of the subiculum-BLA pathway: field potential elicited in BLA before and 40 minutes after delivery of theta burst high-frequency stimulation.

REFERENCES

1. Aggleton, J. P.: A description of intra-amygdaloid connections in the old world monkeys. Exp. Brain. Res., 57, 390-399 (1985).
2. Amaral, D.: Memory: Anatomical organization of the candidate brain regions. In F. Plum (Ed.), Handbook of physiology, sec 1: Neurophysiology, Vol 5: Higher functions of the brain (pp. 211-294). Bethesda, MD: American Physiological Society (1987).
3. Andersen, P., Bliss, T.V.P., Skerde, K. K.: Unit analysis of population spikes. Exp. Brain Res., 13, 208, 221 (1971).
4. Becahra, A., et al.: Double dissociation of conditioning and declarative knowledge relative to the amygdala and hippocampus in humans. Science, 269, 1115-1118 (1995).
5. Bouton, M. E., Bolles, R. C.: Conditioned fear assessed by freezing and by the suppression of three different baselines. Anim Learn Behav, 8, 429-434 (1980).
6. Campeau, S., Davis, M.: Lesions of the auditory thalamus block acquisition and expression of aversive conditioning to an auditory but not a visual stimulus measured with the fear potentiated startle paradigm. Soc. Neurosci. Abstr, 17, 658 (1991).
7. Campeau, S., Miserendino, M. J. D., Davis, M.: Intra-amygdala infusion of the N-methyl-D-aspartate receptor antagonist AP5 blocks acquisition but not expression of fear potentiated startle to an auditory conditioned stimulus. Behav. Neurosci., 106, 569-574 (1992)
8. Chapman, P. F., Bellavance, L. L.: Induction of long-term potentiation in the baso-lateral amygdala does not depend on NMDA receptor activation. Synapse, 11, 310-318 (1992).
9. Chapman, P. F., et al.: Long-term synaptic potentiation in the amygdala. Synapse, 6, 271-278 (1990).
10. Clugnet, M. C., LeDoux, J. E.: Synaptic plasticity in fear conditioning circuits: Induction of LTP in the lateral nucleus of the amygdala by stimulation of the medial geniculate body. J. Neurosci., 10, 2818-2824 (1990).
11. Davis, M.: The role of the amygdala in conditioned fear. In J. P. Aggelton (Ed.), The amygdala: Neurobiological aspects of emotion, memory, and mental dysfunction (pp.255-305). New York: Wiley-Liss (1992)
12. Fanselow, M. S., Kim, J. J., Landeira-Fernandez, J.:

Anatomically selective blockade of Pavlovian fear conditioning by application of an NMDA antagonist to the amygdala and periqaueductal gray. Soc Neurosci Abstr, 17, 659 (1991)
13. Farb, C. F., LeDoux, J. E., Milner, T. A.: Glutamate is present in medial geniculate body neurons that project to lateral amygdala and in lateral amygdala presynaptic terminals. Soc. Neurosci. Abstr., 15, 890 (1989).
14. Gallagher, M, Holland, P. C.: Understanding the function of the central nucleus: Is simple conditioning enough? In J. P. Aggelton (Ed.), The amygdala: neurobiological aspects of emotion, memory, and mental dysfunction (pp.307-322). New York: Wiley-Liss (1992).
15. Hitchcock, J. M., Davis, M. : Lesions of the amygdala but not of the cerebellum or red nucleus, block conditioned fear as measured with the potentiated startle paradigm. Behav. Neurosci., 100, 11-22 (1986).

16. Iwata, J., et al.: Intrinsic neurons in the amygdaloid field projected by the medial geniculate body mediate emotional responses conditioned to acoustic stimuli. Brain Res., 388, 195-214 (1986).
17. Izquierdo, I., et al.: CNQX infused into rat hippocampus or amygdala disrupts the expression of memory of two different tasks. Behav. Neurol. Biol. 59, 1-4 (1993).
18. Kairiss, E. W., et al.: Field potential evidence for LTP or feed forward inhibition in the rat dentate gyrus. Brain Res., 401, 87-94 (1987).
19. Kapp, B. S., et al.: Amygdala central nucleus lesions: Effects on heart rate conditioning in the rabbit. Physiol. Behav. 23, 1109-1117 (1979).
20. Kim, J. J. et al.: N-methyl-D-aspartate receptor antagonist APV blocks acquisition but not expression ofd fear conditioning. Beahv. Neurosci., 105, 126-133 (1991).
21. Kim, J. J. et al.: Infusion of the non-NMDA receptor antagonist CNQX into the amygdala blocks the expression of fear potentiated startle. Behav. Neurol. Biol., 59, 5-8 (1993).
22. Kim, J. J. et al.: Selective impairment of long-term but not short-term conditional fear by the N-methyl-D-aspartate antagonist APV. Behav. Neurosci. 106, 591-596 (1992).
23. Kim, J. J., Fanselow, M. S.: Modality specific retrograde amnesia of fear. Science, 256, 675-677 (1992).

24. LeDoux, J. E., Sakaguchi, A., Reis, D. G.: Subcortical efferents projections of the medial geniculate nucleus mediate emotional responses conditioned to acoustic stimuli. J Neurosci, 4, 683-698 (1984).
25. LeDoux, J. E., et al.: The lateral amygdaloid nucleus: Sensory interface of the amygdala in fear conditioning. J Neurosci, 10, 1062-1069 (1990).
26. Lomo, T.: Patterns of activations in a monosynaptic cortical pathway: the perforant path input to the dentate area of the hippocampal formation. Exp. Brain Res., 12, 18-45 (1971).
27. Maren, S., Fanselow, M. S.: Synaptic plasticity in the basolateral amygdala induced by hippocampal formation stimulation in vivo. J. Neurosci., 15, 7548-7564 (1995).
28. Maroun, M., Yaniv, D., Richter-Levin G.: Local circuit plasticity in the dentate gyrus following high-frequency stimulation. Isr. J. Medic. Sci., 32 sup., 41 (1996).
29. Miserendino, M. J. D., et al.: Blocking of acquisition but not expression of conditioned fear potentiated startle by NMDA antagonists in the amygdala. Nature, 319, 716-718 (1990)
30. Phillips, R. G., LeDoux, J. E.: Differential contribution of amygdala and hippocampus to cued and contextual fear conditioning. Behav. Neurosci., 106, 274-285 (1992a).
31. Phillips, R. G., LeDoux, J. E.: Overlapping and divergent projections of CA1 and the ventral subiculum to the amygdala. Soc. Neurosci. Abstr. 18, 518 (1992b).
32. Smith, B. S., Millhouse, O. E.: The connections between basolateral and central nuclei. Neuroscience letters, 56, 307-309 (1985).

ROLE OF NITRIC OXIDE ON EMOTIONAL AND MOTOR BEHAVIOUR

E. A. Del Bel[1], R.M.W. de Oliveira[2], F. S. Guimarães[2].
Depts. of [1]Physiology, FORP and [2]Pharmacology, FMRP, Campus USP, Ribeirão Preto, SP, 14049-900, Brazil.

Abstract: Nitric Oxide (NO) has been proposed as a new neurotransmitter/neuromodulator in the Central Nervous System (CNS). Microinjections of inhibitors of NO synthase (NOS), the enzyme responsible for NO formation, into the dorsal central grey, induce anxiolytic effects in rats. On the other hand, SIN-1, a NO donor, microinjected into these regions, induces flight reactions. However, although systemic injections of low doses of NOS inhibitors induce anxiolytic effects in the elevated plus maze, higher doses decrease maze exploration. This may involve motor effects of this compounds, since we showed that they can induce catalepsy in a dose dependent manner in rodents either after systemic or Intra-striatal injections. These results suggest that NO may be related to the control of motor and emotional behaviour.
Key words: nitric oxide, anxiogenic, anxiolytic, catalepsy, L-NOARG, elevated plus maze, l-arginine, haloperidol, tolerance, Dopamine, NADPH diaphorase, intracerebral injection, dorsal central grey.

Nitric oxide (NO) is a short lived, highly reactive molecule, postulated to act as a neuromodulator or second messenger molecule in neural and non neural systems [7,10,17,32,50]. NO is synthesised from L-arginine by the enzyme Nitric Oxide Synthase (NOS), that exist in several forms [16]. The brain enzyme is constitutive, cytosolic, Ca++/Calmodulin dependent [8], occurs only in neuronal cell bodies, dendrites and axons [7], and presents discrete location in brain structures such as the cortex, cerebellum, hypothalamus, caudate and hippocampus [4,54]. The samecerebral neurons stain for NOS and NADPH diaphorase [12,23], and purified brain NOS has NADPH diaphorase activity [8].

Nitric Oxide plays a role in a variety of biological mechanisms. Manipulations of NO formation modify many brain physiological and/or pathological conditions such as long-term potentiation in hippocampus [5,34], long-term depression in cerebellum [25,47], neurotoxicity/neuroprotection [10,27], epilepsy [14,30], chronic actions of opiates [24], nociception [2,3,11] and anxiety [15,21,51]. Among regions with significant levels of NOS/NADPH-diaphorase are sites involved in modulation of defensive responses such as the amygdala, hypothalamus and dorsolateral central grey (DCG, figure 1). These regions have been related to elaboration of flight/fight reactions to threatening stimuli [18,19].

We showed that microinjection of L-NAME or L-NOARG, two inhibitors of NO formation, into the dorsal central grey (DCG), increased open arm exploration of an elevated plus maze, without affecting the number of enclosed arm entries, indicating an anxiolytic effect of both compounds [21]. The effect of L-NAME was prevented by previous treatment with l-arginine, suggesting it is related to the inhibition of NOS (figure 2). However, the range of effective doses was limited, and the dose-effect curves had an inverted U shape, higher doses tending to reduce maze exploration.

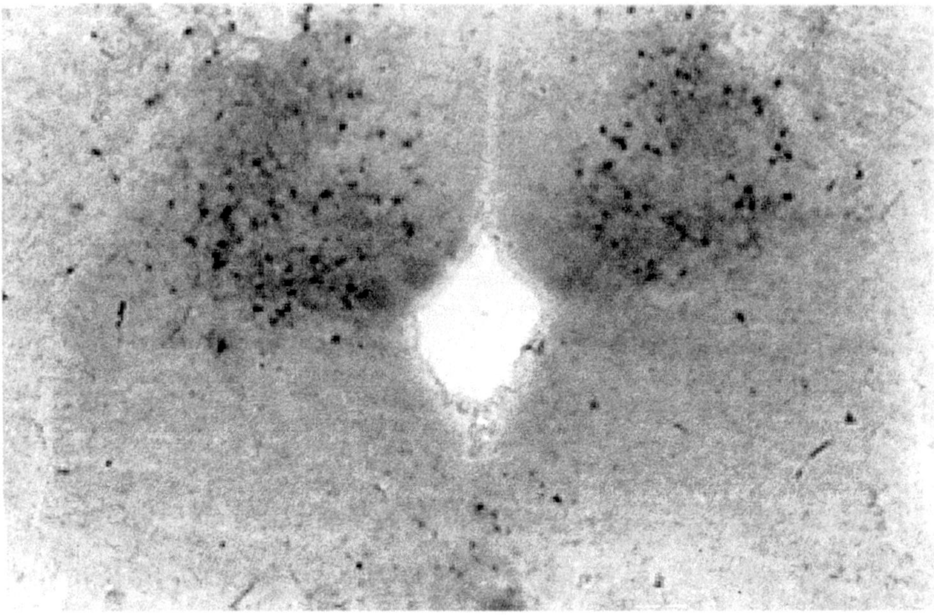

Figure 1. NADPH-diaphorase/NOS positive cells located in the dorsolateral part of the PAG. The reaction for NADPH-diaphorase was performed using the method employed by Vincent and Kimura [54], with modifications. Rats were anethesized i.p. with 25% urethane and perfused through the left ventricle with 50 ml of 0.9% saline followed by 100 ml 4% paraformaldehyde, and 0.1 M phosphate buffer solution. The brains were removed, soaked for 2 h in 4% paraformaldehyde, 0.1 M phosphate buffer solution and overnight in 30% sucrose 0.1 M phosphate buffer solution, frozen in liquid nitrogen, and stored at -70°C. Thirty µm coronal brain sections were then cut with a cryostat. NADPH-diaphorase activity was demonstrated by incubating sections (free-floating) in 0.1 M phosphate buffer, pH 7.4, containing 0.3% Triton-X100, 0.1 mg/ml nitrobluetetrazolium and 1.0 mg/ml ß-NADPH at 37 °C for 30-60 min in the dark. The sections were then dipped for 10 min in phosphate buffer, dried, rinsed in distilled water, dried again and slices were mounted on slides for microscopic observation.

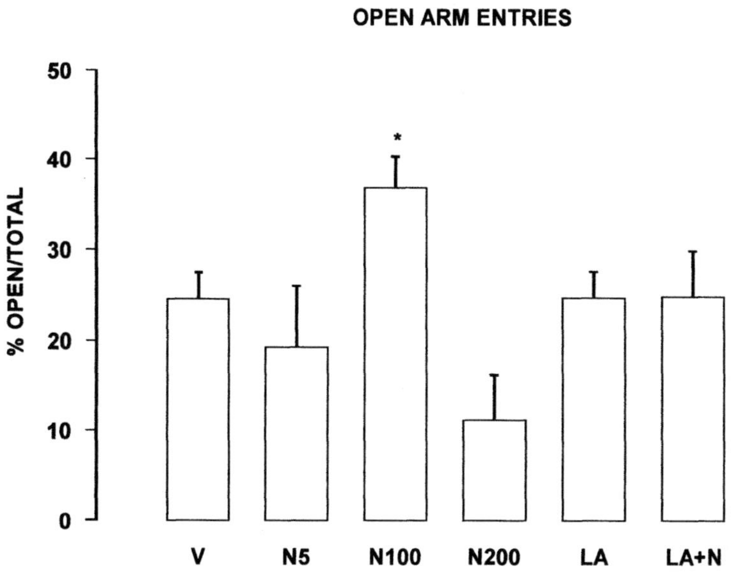

Figure 2. Anxiolytic effect of L-NAME on the elevated plus maze. Male Wistar rats (200-250 g), with implanted cannulas into the dorsal central grey (DCG, coordinates: 1.9 mm lateral to the right side of lambda at an angle of 22° with the sagital plane), received, seven days after surgery, microinjections of into this region of L-NAME (N, 5 to 200 nmol) or vehicle (V) and were submitted to the elevated plus maze, an animal model of anxiety [20], for 5 min. In this test, exploration indices of open arms are considered to be inversely related to anxiety. A second group of animals received microinjections of L-arginine (50 nmol) + vehicle (LA) or L-arginine + L-NAME (100 nmol, LA + N). Bars represent the means ± SEM. Asterisks indicate significant difference from vehicle.

To further investigate a possible role of NO on anxiety modulation in the DCG, we decided to verify the behavioural effects of SIN-1, a NO donor [32]. Animals (N=3-4/group) received microinjection into the DCG of saline or SIN-1 (150 or 300 nmol) and were immediately placed in the centre of an open arena (diameter: 60 cm, surrounded by a Plexiglas wall 35 cm high). Behaviour in the arena was recorded by a video-camera during 10 min. Preliminary results showed that all but one of the rats receiving SIN-1 displayed defensive reactions, characterised by wild running bursts, defecation and vertical jumps oriented to the edge of the apparatus, leading to effective escape in two cases. The reaction subsides after approximately 10 min. We also studied, through detection of the proto-oncogene c-Fos, neuronal areas activated after microinjection of SIN-1 into this region. The animals were sacrificed 2 or 5 h later and had their brains removed and processed for Fos-like immunoreactivity (FLI) detection.

Neurons with FLI were found in several brain regions of saline treated animals, probably as a consequence of handling and/or exposure to the open arena. Animals showing defensive reactions after SIN-1, however, had significant increases in the number of cells with FLI amygdaloid nuclei (83 and 376%, respectively, $F_{3,4}=9.0$, $P=0.029$), diagonal band ($F_{3,7}=20.2$, $P=0.0008$, 217 and 228%), anterior hypothalamus (708 and 780%, $F_{3,5}=15.6$, $P=0.006$) and lateral hypothalamus (790%, $F_{3,6}=5.5$, $P=0.037$), as compared to saline. There was also a trend for an increase in FLI in the cingulate and piriform cortex, medial hypothalamus, dorsal raphe nucleus and central grey. No difference was found in the number of FLI neurons of rats sacrificed 5 h after SIN-1 administration, as compared to saline ($P>0.10$, figures 3 and 4).

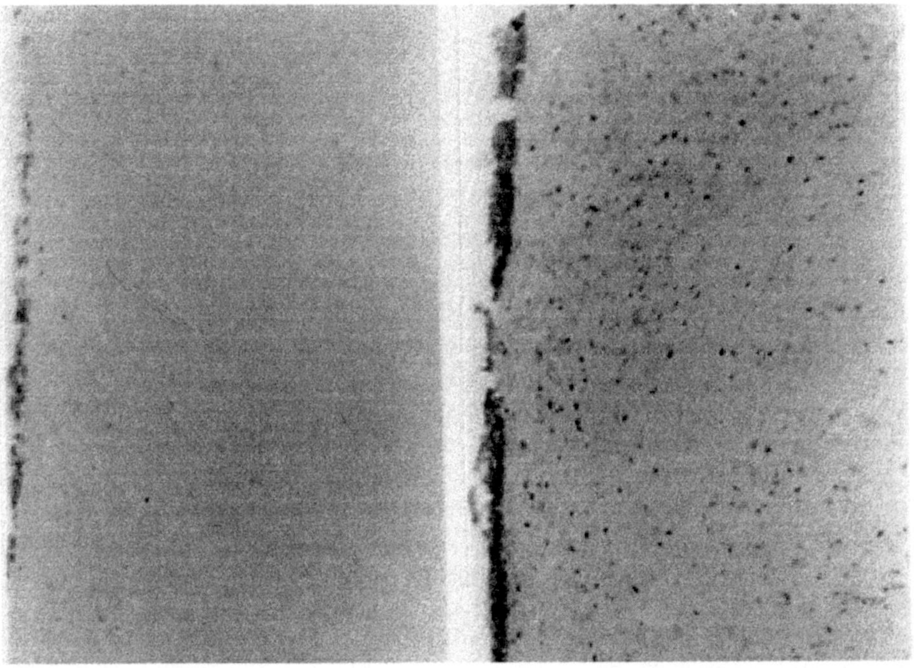

Figure 3. Fos-like immunoreactivity (FLI) in the medial hypothalamus of rats receiving microinjections into the DCG of SIN-1 (300 nmol, right panel) or saline (left panel). Two h after the injections the animals were anaesthetised with i.p. injections of urethane 25% (2.5 ml/kg) and transcardiacly perfused with saline followed by 4% paraformaldehyde. After being removed, the brains were kept overnight in 30% sucrose solution. Afterwards, 35 μm coronal sections were obtained in a cryostat and processed for Fos immunohistochemistry, as described by Silveira et al. [49]. Briefly, tissue sections were successively washed and incubated for 40 h with the primary Fos antibody (1:2000, Cambridge Research Biochemicals). Sections were then processed by the avidin-biotin immunoperoxidase method (Vectastain ABC kit, Vector Lab.) and FLI was

revealed by addition of the chromogen 3,3'-diaminobenzidine (DAB, Sigma) and hydrogen peroxide. The slices were mounted on slides and cover slipped for microscopic observations. FLI could be visualised as a brown reaction product inside neuronal nuclei, whose location was determined by use of Paxinos and Watson's atlas [39].

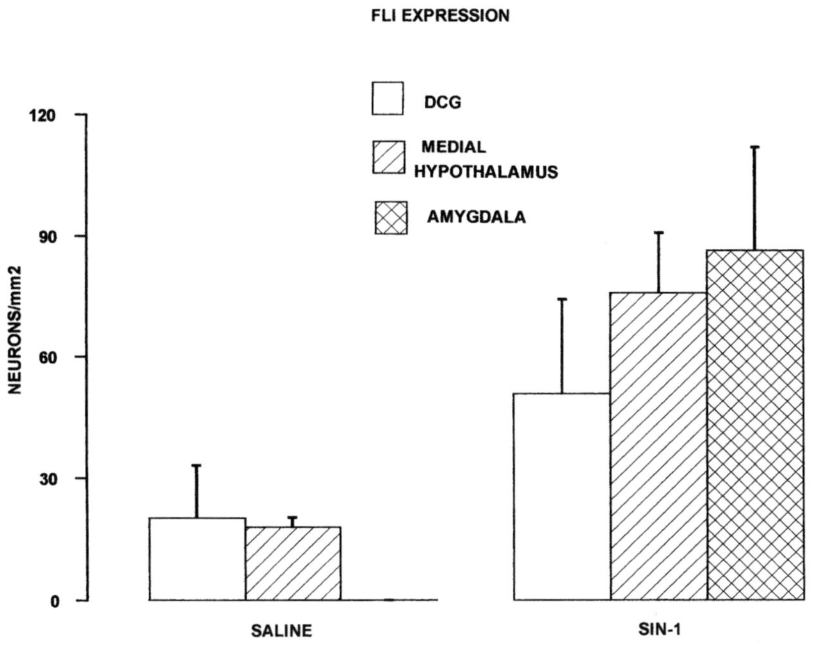

Figure 4. Fos-like immunoreactivity (FLI) in the dorsal central grey (DCG), medial hypothalamus and amygdaloid nuclei of rats receiving microinjections into the DCG of SIN-1 (300 nmol). The number of FLI neurons in selected regions was automatically recorded by JAVA computerised image analysis system. Bars represent the means ± SEM.

The neurochemistry of the DCG has been investigated by several authors (for review, see Graeff, [19]). It is well documented that gabaergic neurons play a tonic inhibitory role in this region. Glutamate, on the other hand, is able to stimulate the DCG through both NMDA and non-NMDA receptors [19,20,29]. Accordingly, both biccuculine, a GABA-A receptor antagonist, and kainic acid, a non-NMDA glutamate receptor agonist, induce flight reactions when microinjected into the DCG [19].

NO has been proposed as a transduction mechanism of NMDA receptor-mediated neurotransmission, at least in some regions of the central nervous system. Activation of NMDA receptors leads to formation of NO which, in turn, seems to mediate NMDA-induced increase in intracellular cGMP [17]. However, the effects of NO on NMDA mediated neurotransmission are now recognised as being more complex than initially thought. Besides increasing cGMP, NO can also inhibit NMDA-induced intracellular Ca++ entry and NOS activity, exercising a local negative feedback [31]. More recently, Lipton et al. [27] showed that NO can also antagonise NMDA receptors.

The flight reaction induced by SIN-1 differ significantly from that usually seem after biccuculine and kainic acid [19]. Different from the latter, there are escape attempts, with jumps towards the edge of the Plexiglas wall. This pattern of reaction resembles those seem after electrical or chemical stimulation of the hypothalamus [19]. It is interesting, therefore, that following the flight reaction induced by SIN-1 in the DCG there was a large increase in FLI in many hypothalamic nuclei. FLI in the hypothalamus was even greater than that seen in the DCG. Actually, in the latter there was no significant difference from saline injected groups. Although it must be considered that saline group had a greater than normal FLI neurons, probably because the animals were also exposed to the open arena, a bright new environment with aversive properties, there seems to be a relation between the pattern of flight behaviour displayed by the animals (with escape attempts) and the large number of FLI found in hypothalamic nuclei.

The mechanism involved in such effect is unknown. NO can diffuse to other neurons, increasing presynaptic glutamate release [31]. Consequently, it has been proposed to act as a 'retrograde' messenger [31]. It is possible, then, that it could be retrogradely activating descending pathways from the hypothalamus to the DCG. More studies, however, are needed to address this issue.

Studies with systemically injected drugs have produced contradictory results on the participation of the L-Arginine/NO pathway in anxiety. For example, Quock and Nguyen [40] and Caton et al. [9] found that L-NOARG decreases the anxiolytic effect of nitrous oxide and chlordiazepoxide in mice, suggesting an anxiolytic action for NO. Also, L-NOARG treatment aggravates gastric ulcers induced by stress [35] and disrupts emotional habituation based on defecation scores [38]. In contrast, other reports point to an anxiogenic action of NO in the CNS of rats. Volke et al. [55] and Faria et al. [15] have shown that L-NAME has an anxiolytic effect in the elevated plus maze. In the former study, however, doses higher than 10 mg/kg were not effective.

To further investigate the role of NO in anxiety, we [26] verified the effects of L-NOARG (5-120 mg/kg i.p.) in the elevated plus maze. Acute systemic injections of the drug in rats decrease open arm exploration in the plus maze, interpreted as an anxiogenic effect (Figure 5). The acute effect of L-NOARG was attenuated dose-dependently by i.c.v pre-treatment with l-arginine. These findings suggest that L-NOARG effect in the elevated plus maze involves a decrease in NO formation in the CNS.

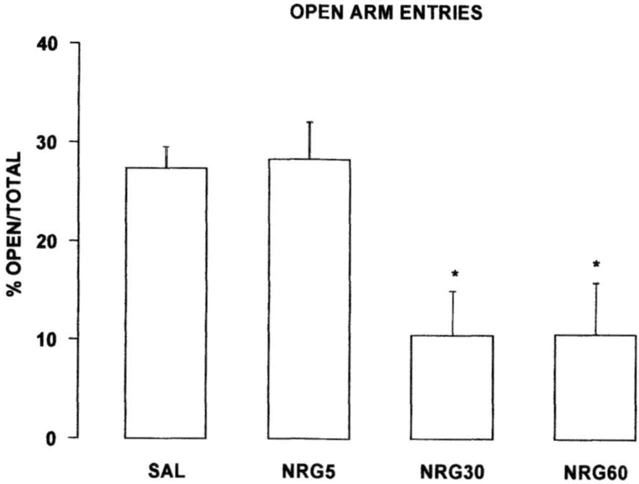

Figure 5. Effect of systemic injections of L-NOARG on open arm exploration of an elevated plus maze. Male Wistar received i.p. injection of L-NOARG (NRG, 5-60 mg/kg) and were tested in the plus maze 1 h after. Further specifications as in figure 2.

The decreased maze exploration induced by acute L-NOARG treatment totally disappeared after four days of L-NOARG administration, a rapid tolerance. A confounding element in the interpretation of results obtained with NOS inhibitors is the drug effects on blood pressure. Acute administration of L-NOARG causes an increase in systemic blood pressure [41]. Some studies have suggested an influence of blood pressure on anxiety [36]. Nevertheless, although increased levels of blood pressure are maintained after four days of L-NOARG treatment [36], the drug is no longer able to modify exploratory indexes in the elevated plus maze. In addition, Faria et al. [15] describe that two-Kidney one-clip hypertension failed to change the behaviour of rats in the elevated plus maze. In the same way, Rosa et al., [42] showed that hypertension per se does not affect the level of fear in spontaneously hypertensive rats submitted to the elevated plus maze. Therefore, it seems unlikely that increased blood pressure, per se, could explain the effect of acute L-NOARG administration.

In our study the drug also decreased the number of enclosed arm entries, an index usually related to general exploratory activity (Figure 6). Therefore, the "anxiogenic" effect obtained in our study should be interpreted with caution since the doses employed were comparatively higher than the active doses of L-NAME that showed anxiolytic effects in other studies [15,55]. A possible interference of NOS inhibitors on motor behaviour could help to explain the inverted U shaped dose-response curves found in several studies [21,55].

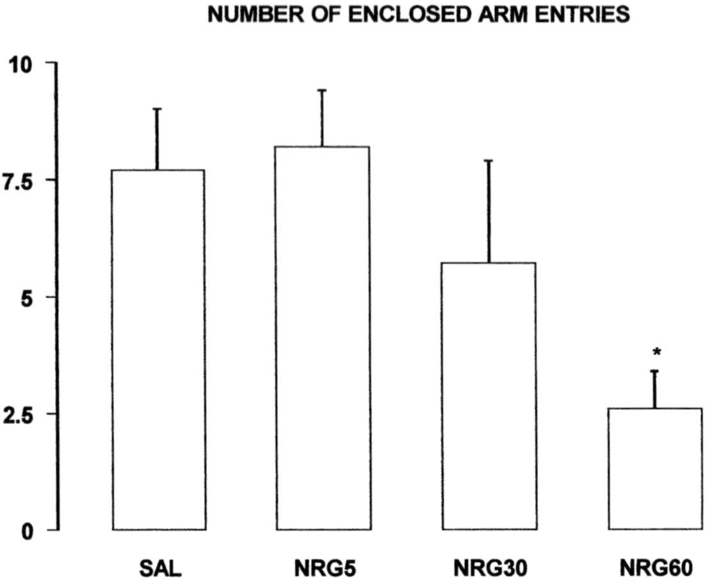

Figure 6. Effect of systemic injections of L-NOARG on the number of enclosed arm entries of an elevated plus maze. Further specifications as in figure 5.

There are NADPH-diaphorase/NOS neurons in the cortex, striatum, pedunculopontine tegmental nucleus [54], regions involved in motor control (Figure 7). In addition, inhibition of NO formation decreases dopamine release in the striatum [46] and antagonises the increase in locomotor activity found after dopamine agonists administration [52]. Based on these findings we decided to test if NOS inhibitors would potentiate haloperidol induced catalepsy in mice. Catalepsy is defined as a failure to correct an externally imposed posture. This test is widely used to evaluate motor effects of drugs, particularly those related to the extrapyramidal system. Drugs such as haloperidol, that antagonise dopamine neurotransmission in the striatum, induce catalepsy in rodents and Parkinson symptoms in humans [44]. We found that systemic injections of L-NOARG induce catalepsy in mice dose-dependently (Figure 8); it actually had an additive effect with haloperidol, that also caused catalepsy per se [13,28]; the effect was attenuated by previous treatment with L-Arginine, and suffered tolerance after four days of chronic (twice a day) administration. The mechanisms involved in catalepsy induced by NOS inhibition and in the rapid development of tolerance are not clear. It does not involve changes in blood pressure since blood pressure levels are still elevated after 4 days treatment [41].

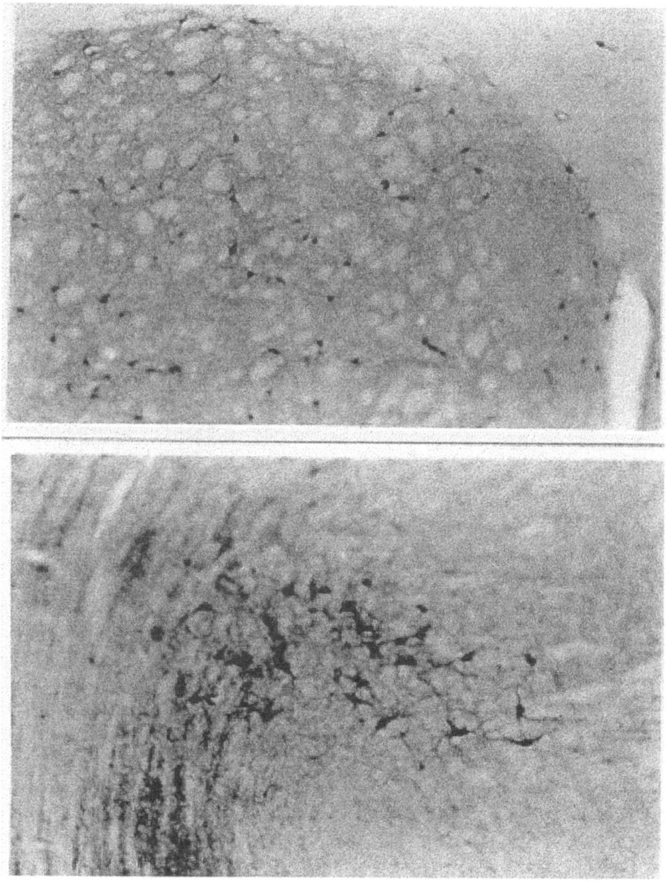

Figure 7. NADPH-diaphorase/NOS positive cells located in the striatum (upper panel) and pedunculopontine tegmental nucleus (lower panel). Further specifications as in figure 1.

L-NOARG induced catalepsy probably involves a decrease in NO formation in the CNS. The doses of inhibitors employed in the experiments were within those used in other studies [45,52] and were similar to those shown by Salter et al. [43] to significantly inhibit neuronal NOS (> 10 mg kg^{-1}).

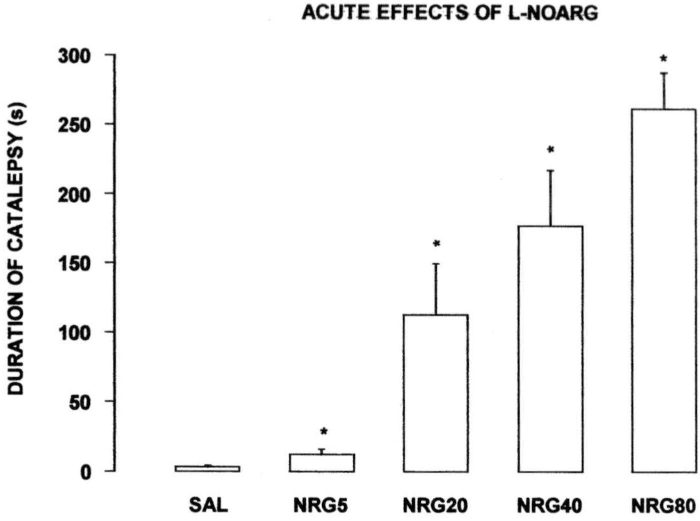

Figure 8. Catalepsy induced by L-NOARG (NRG, 5-80 mg/kg, n=8-10 mice/group) administered i.p. 1 h before the test. Catalepsy was evaluated by placing the animal with both forelegs over a horizontal glass bar (diameter 0.5 cm) elevated 4.5 cm from the floor. The time (in sec) during which the mouse maintained this position was recorded up to 300 s. Bars represent the means ± SEM. Asterisks indicate significant difference from saline (ANOVA followed by the Duncan test, $P < 0.05$).

Besides, intracerebro-ventricular and striatum microinjection of NOS inhibitors produces catalepsy in rats (figure 9 and 10). Therefore, although contrary results also exist [48], it is possible that L-NOARG decreases locomotor activity by interfering with striatal systems, probably with dopaminergic neurotransmission. There are several evidences that support this possibility. Animals that become tolerant to the motor effects of L-NOARG also become tolerant to haloperidol (Figure 11). Antagonism of NO formation attenuates dopamine release in the striatum [6,46] and inhibits the increased locomotor activity found after dopamine agonists administration [1,52]. Several studies have shown that antagonism of NMDA-mediated transmission attenuates catalepsy induced by dopamine receptor antagonists such as haloperidol [33,38,56]. Since animals tolerant to L-NOARG are tolerant to haloperidol effects, and NO has complex effects on NMDA mediated-neurotransmission, it is also possible that an influence of NO on dopamine neurotransmisson is mediated by effects on NMDA neurotransmission.

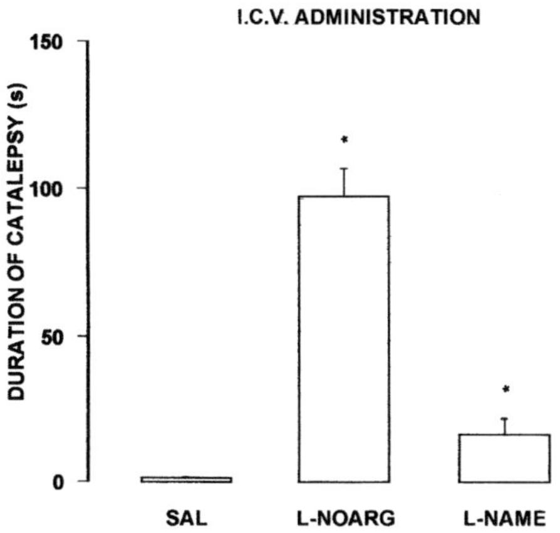

Figure 9. Catalepsy induced by L-NOARG (100 nmol) or L-NAME (100 nmol) microinjected i.c.v. in rats (N=8-10 animals/group). Catalepsy was evaluated by placing the animal with both forelegs over a horizontal glass bar (diameter 0.5 cm) elevated 9.0 cm from the floor. The time (in sec) during which the rat maintained this position was recorded up to 180 s. Bars represent the means ± SEM. Asterisks indicate significant difference from saline (ANOVA followed by the Duncan test, $P < 0.05$).

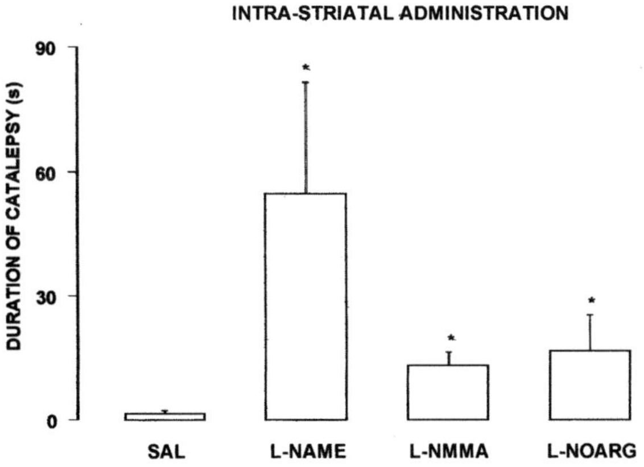

Figure 10. Catalepsy induced by L-NOARG (100 nmol), L-NMMA (100 nmol) or L-NAME (100 nmol) microinjected into the striatum of rats (N=8-10 animals/ group). Further specifications as in figure 10.

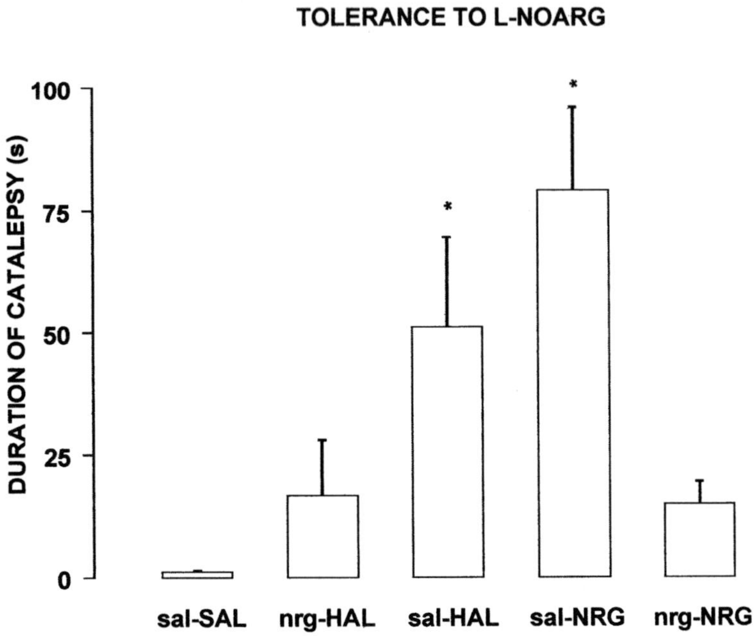

Figure 11. Effects of chronic treatment with saline (sal) or L-NOARG (nrg, 40 mg/kg i.p., twice a day, during 4 days) on catalepsy induced by haloperidol (HAL, 1.0 mg/kg i.p.), or L-NOARG (NRG, 40 mg/kg i.p.). Animals were tested 1 h after the last drug injection. Asterisks indicate significant differences from nrg-HAL and nrg-NRG groups. Further specifications as in figure 8.

Anticholinergic drugs readily reverse neuroleptic induced catalepsy in animals. Atropine sulphate blocks L-NOARG induced catalepsy in mice; this effect is probably centrally mediated since no influence on NOS induced catalepsy was seen after i.p. injection of quaternary atropine (atropine methylnitrate), a compound that does not cross the blood-brain barrier (Figure 12). It is interesting that one of the principal group of cholinergic neurons is the nucleus tegmental pedunculopontinus (NPP) [22]. The NPP cholinergic cell group projects to the extrapyramidal motor system and virtually all cholinergic cells in this region express NOS mRNA [53].

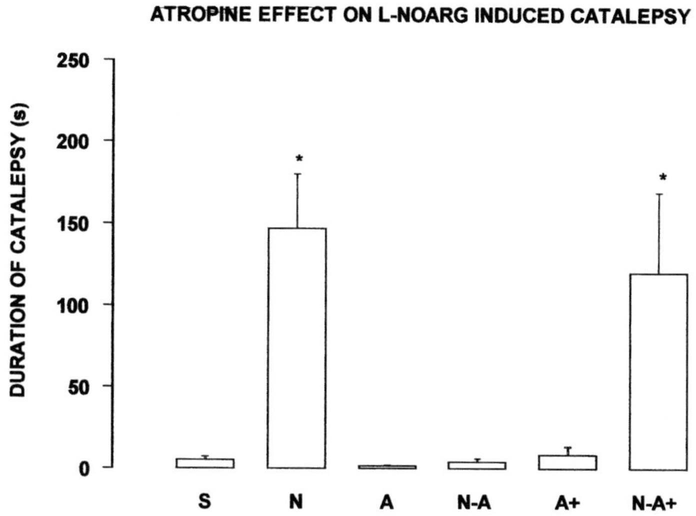

Figure 12. Reversal of L-NOARG (N, 40 mg/kg i.p.) induced catalepsy by atropine (A, 5 mg/kg i.p.). Animals (n=8-10 animals/group) were tested 1 h after the last drug injection. Additional groups received quartenary atropine (atropine methylnitrate, A+, 50 mg/kg i.p.). Asterisks indicate significant difference from animals receiving saline (S). Further specifications as in figure 8).

Conclusions: Our results suggest that NO may modulate emotional reactions in the dorsal central grey. They also suggest that inhibition of NO formation induces significant motor behaviour, probably by opposing dopaminergic neurotransmission in the striatum. These effects may interfere with the results of systemically administered NOS inhibitors compounds in behavioural tests such as the elevated plus maze.

Acknowledgements: We acknowledge the helpful technical support provided by J.C. de Aguiar, C. A da Silva, E.T. Gomes and R. Caldo. The authors were recipients of CNPq and CAPES (RWO) fellowships. The study was supported by grants from CNPq, Finep and FAPESP.

References

1. Abekawa, T.;,Ohmori, T. , Koyama, T.: Effect of NO synthase inhibition on behavioral changes induced by a single administration of methamphetamine. Brain Res. 666, 147-150 (1994)
2. Babbedge, R.C., Hart, S.L., Moore, P.K.: Antinociceptive activity of nitric oxide synthase inhibitors in the mouse: dissociation between the effect of L-NAME and L-NMMA. J. Pharmacol.,45, 77-79 (1993a)

3. Babbedge, R.C. et al.: L-NG-nitro-arginine p-nitroanilide (L-NAPNA) is antinociceptive in the mouse. Neuroreport,4, 307-310 (1993b).
4. Barjavel, M.J, Bhargava, H.N. : Nitric oxide synthase activity in regions and spinal cord of mice and rats: kinetic analysis. Pharmacology 50, 168-174 (1995)
5. Bohme, C.A. et al.: Possible involvement of nitric oxide in long-term potentiation. Eur. J. Pharmacol. 199, 379-381 (1991)
6. Bowyer, J.F., et al.: Nitric Oxide regulation of metjanphetamine-induced dopamine release in caudate / putamen. Brain Res. 699, 62-70 (1995)
7. Bredt, D.S., Hwang, P.M., Snyder, S.H.: Localization of nitric oxide synthase indicating a neural role for nitric oxide. Nature 347, 768-770 (1990)
8. Bredt, D.S., Snyder, S.H.: Isolation of nitric oxide synthetase, a calmodulin-requiring enzyme. Biochemistry 87, 682-685 (1990)
9. Caton, P.W., Tousman, S.A., Quock, R.M.: Involvement of nitric oxide in nitrous oxide anxiolysis in the elevated plus-maze. Pharmacol. Biochem. Behav. 48, 689-692 (1994)
10. Choi, D.W.: Nitric oxide: foe or friend to the injured brain. Proc. Natl. Acad. Sci. USA 90, 9741-9743 (1993)
11. Coderre, T.J.: The role of excitatory amino acid receptors and intracellular messengers in persistent nociception after tissue injury in rats. Mol. Neurobiol. 7, 229-246 (1994)
12. Dawson, T.M. et al.: Nitric oxide synthase and neuronal NADPH diaphorase are identical in brain and peripheral tissue. Proc. Natl. Acad. Sci. 88, 7797-7801 (1991)
13. Del Bel, E.A., da Silva, C.A., Guimarães, F.S.: Catalepsy induced by nitric Oxide Syntase inhibitors. General Pharmacology, in press.
14. Del Bel, E.A. et al.: Anticonvulsant and proconvulsant rolos of Nitric Oxide in experimental epilepsy models Brazilian J. Med. Biol. Res., in press.
15. Faria.M.S. et al.: Acute inhibition of nitric oxide synthesid induces anxiolysis in the plus maze test Europ. J. Pharmacol. 323, 37-43 (1997)
16. Forstermann, U. et al.: Characterization and purification from different cell types. In: Commentary Isoforms of Nitric Oxide Synthase. Biochem. Pharmacol. 42 , 1849-1857 (1991)
17. Garthwaite, J.: Glutamate, nitric oxide and cell-cell signalling in the nervous system. Trends Neurosci. 14, 60-67 (1991)
18. Graeff, F.G. et al.: . Role of 5HT in stress, anxiety and depression. Pharmacol. Biochem. Behav. 54, 129-141 (1996)
19. Graeff, F.G.: Brain defence system and anxiety. In: Roth, M., Burrows, G.D., Noyes, R., (eds) Handbook of anxiety, Vol. 3. Amsterdam: Elsevier, 1990, pp. 307-354
20. Guimarães, F.S. et al.: Anxiolytic effect in the elevated plus-maze of the NMDA receptor antagonist AP7 microinjected into the dorsal periaqueductal grey. Psychopharmacology 103, 91-94 (1991)
21. Guimarães, F.S. et al.: Anxiolytic effect of nitric oxide inhibitors microinjected into the dorsal central grey. NeuroReport 5, 1929-1932 (1994)

22. Hirsh, E. et al.: Neuronal loss in the pedunculopontine tegmental nucleus in Parkinson disease and progressive supranuclear palsy. Proc. Natl. Acad. Sci. USA 84, 5976-5980 (1987)
23. Hope, B.T. et al.: Neuronal NADPH diaphorase is a nitric oxide synthase. Proc. Natl. Acad. Sci. USA 88, 2811-2814 (1991)
24. Kolesnikov, Y.A., Pick, C.G., Pasternak, G.W.: N^G-Nitro-L-Arginine prevents morphine tolerance. Eur. J. Pharmacol. 221, 399-400 (1992)
25. Linden, D.J., Connor, J.A.: Long-term depression of glutamate currents in cultured cerebellar Purkinje neurons does not require nitric oxide signalling. Eur. J. Neurosci. 4, 10-15 (1991)
26. Lino de Oliveira, C., Del Bel, E.A., Guimarães, F.S.: Effects of L-NOARG on plus-maze performance in rats. Pharmacol. Biochem. Behav. 56, 55-59 (1997)
27. Lipton, S.A. et al.: A redox-based mechanism for the neuroprotective effects of nitric oxide and related nitroso-compounds. Nature 364, 626-632 (1993)
28. Marras, R.. et al.: L-NOARG, an inhibitor of nitric oxide synthase, induces catalepsy in mice. NeuroReport 7, 158-160 (1995)
29. Matheus, M.G., Guimarães, F.S.: Antagonism of non-NMDA receptors in the dorsal periaqueductal grey induces anxiolytic effect in the elevated plus maze. Psychopharmacology, in press.
30. Mollace, V., Bagetta, G., Nistico, G.: Evidence that L-arginine possesses proconvulsant effects mediated through nitric oxide. Neuroreport 2, 269-272 (1991)
31. Moncada, S., Higgs, A.: The l-arginine pathway. N. Engl. J. Med. 329, 2002-2012 (1993)
32. Moncada, S., Palmer, R.M.J., Higgs, E.A.: Nitric oxide: physiology, pathophysiology, and pharmacology. Pharmacol. Rev. 43, 109-142 (1991)
33. Moore, N.A. et al.: NMDA receptor antagonism inhibit catalepsy induced by either dopamine D1 or D2 receptor antagonist. Eur. J. Pharmacol. 237, 1-7 (1993)
34. O'Dell, T.J. et al.: Test of the roles of two difusible substances in long-term potentiation: evidence for nitric oxide as a possible early retrograde messenger. Proc Natl. Acad. Sci. USA 88, 11285-11289 (1991)
35. Ogle, C.W., Qiu,B.S.: Nitric oxide inhibition intensifies cold-restraint induced gastric ulcers in rats. Experentia 49, 304-307 (1993)
36. Oliveira, G.M., Rosa, W.C.M., Nakamura-Palacios, E.M.: Effect of age and magnesium administration on the performance of SHR in the elevated plus-maze. Brazilian J. Med. Biol. Res. 28, 343-346 (1995)
37. Onstott, D., Mayer, B., Beitz, A.J.: Nitric oxide synthase immunoreactive neurons anatomically define a longitudinal dorsolateral column within the midbrain periaqueductal grey of rat: analysis using laser confocal microscopy. Brain Res. 610, 317-324 (1993)
38. Papa, M., Pellicano, M.P., Sadile, A.G.: Nitric oxide and long-term habituation to novelty in the rat. Ann. N. Y. Acad. Sci. 738, 316-324 (1993)
39. Paxinos, G., Watson, C.: The rat brain in stereotaxic coordinates. second edition, San Diego: Acdemic Press, 1986

40. Quock, R.M., Nguyen, E.: Possible involvement of nitric oxide in chlordiazepoxide-induced anxiolysis in mice. Life Sci. 51, 255-260 (1996)
41. Ribeiro, M.O. et al.: Chronic inhibition of nitric oxide synthesis: a new model of arterial hypertension. Hypertension 20, 298-303 (1994)
42. Rosa, W.C.M., Oliveira, G.M., Nakamura-Palacios, E.M.: Effects of antihypertensive drugs alfa-methyldopa and hydralazine on plus maze. Braz.J. Med. Biol. Res. 27, 55 (1994)
43. Salter, M., Duffy, C., Hazelwood, R.: Determination of brain nitric oxide synthase inhibition in vivo: Ex vivo assays of nitric oxide synthase can give incorrect results. Neuropharmacology 34, 327-334 (1995)
44. Sanberg, P.R. et al.: The catalepsy test: its ups and downs. Behav. Neurosci. 102, 748-759 (1988)
45. Sandi, C., Venero, C., Guaza,C.: Decreased spontaneous motor activity and startle response in nitric oxide synthase inhibitor-treatted rats Eur. J. Pharmacol. 277, 89-97 (1995)
46. Sandor, N.T. et al.:Role of nitric oxide in modulating neurotransmitter release from rat striatum. Brain Res. Bull. 36, 483-486 (1995)
47. Shibuki, K., Okada, D.: Endogenous nitric oxide release required for long term synaptic depression in the cerebellum. Nature 349, 326-328 (1991)
48. Silva, M.T. et al.: Increased striatal dopamine efflux in vivo following inhibition of cerebral nitric oxide synthase by the novel monosodium salt of 7-nitro indazole. Brit. J. Pharmacol. 114, 257-258 (1994)
49. Silveira, M.C.L. et al.:C-fos immunoreactivity in the brain following electrical or chemical stimulation of the medial hypothalamus of freely moving rats. Brain Res. 674, 265-274 (1995)
50. Snyder, S.H., Bredt, D.S.: Nitric oxide as a neuronal messenger. Trends Pharmac. Sci. 12, 125-128 (1991)
51. Spinella, M., Bodnar, R.J.: Nitric oxide synthase inhibition selectively potentiates swim stress antinociception in rats. Pharmacol. Biochem. Behav. 47, 727-733 (1994)
52. Star, M.S., Star, B.S.: Do NMDA receptor-mediated changes in motor behaviour involve nitric oxide? Europ. J. Pharmacol. 272, 211-217 (1994)
53. Sugaya, K., McKiney, M.: Nitric Oxide Synthase expression in cholinergic neurons in the rat brain examined by combined immunocythochemistry and in situ hybridization histhochemistry. Mol. Brain Res. 23, 111-125 (1994)
54. Vicent, S.R., Kimura, H.: Histochemical mapping of nitric oxide synthase in the rat brain. Neuroscience 46, 755-784 (1992)
55. Volke, V. et al.: Inhibition of nitric oxide synthase causes anxiolytic-like behaviour in an elevated plus-maze. NeuroReport 6, 1285-1288 (1995)
56. Yoshida, Y. et al.: Distinct sites of dopaminergic and glutamatergic regulation of haloperidol-induced catalepsy within the rat caudadte-putamen. Brain Res. 639, 139-148 (1994)

Multi-Site Recording of Neural Activity using Planar Electrode Arrays

Y. Jimbo and A. Kawana
NTT Basic Research Laboratories, Kanagawa 243-0198, JAPAN.

Keywords. electrode array, cell culture, slice, spike, field potential

1 Introduction

Since the first proposal for an Ion-Sensitive FET in 1970 [4], micro-fabrication technology has become one of the key-techniques for building sensor devices for biological measurements. With technical innovation in the field of electronic engineering, sensors for neurophysiological recording have also been investigated extensively. The most promising advantage of this type of sensor is the capability for integrating multiple electrodes on a single device. This has allowed us to visualize spatially propagating activity at a fine time resolution.

These sensors are categorized roughly into three types. One is referred to as a probe-type device. Multiple electrodes as well as signal processing circuits, such as multiplexors, are integrated directly on semiconductor substrates [19]. Needle-shaped devices have been developed and used to carry out *in vivo* recording from cortical tissue [3]. The second type is the regeneration electrode [17]. Fine through holes are formed on semiconductor substrates using three-dimensional fabrication techniques and the edges of the holes are metallized. This device is implanted at the site of damaged peripheral nerve bundles and recording and stimulation can be achieved after regenerating neurites grow through the holes. The last type, planar electrode array is mainly used for *in vitro* measurements. These consist of a kind of culture dish with embedded multiple microelectrodes. Dissociated neurons or slices are cultured on the dishes. Both recording and stimulation become possible after establishment of electrical contacts between the cultured tissue and the substrate electrodes.

In the following sections, first, basic properties of the planar electrode arrays are briefly described. Then, two examples of measurements with this type of device, stimulation and recording from dissociated neurons and recording from cultured slices, are shown.

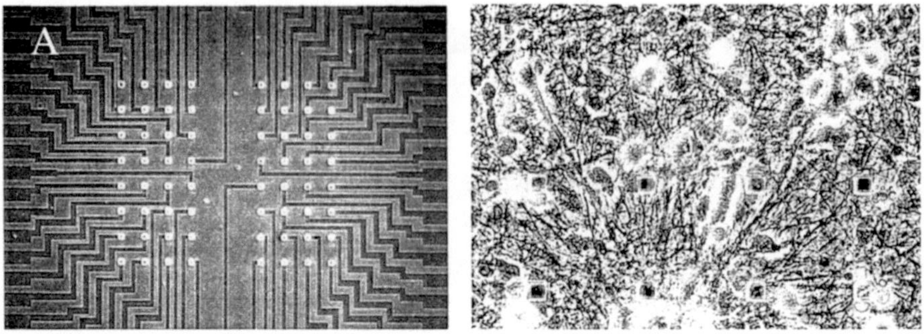

Figure 1 A planar electrode array. **A**. The recording area consists of two separated regions, each of which contains 32 recording sites. The size of each recording site is 30 x 30 μm and the distance is 180 μm. **B**. Dissociated cortical neurons cultured on the planar electrode array. The dark squares are the recording electrodes.

2 Planar Electrode Array

Planar electrode arrays were developed by Gross for spike recording from spinal cord cultures [9]. Since then, the method has been applied to a wide variety of tissues, such as dissociated invertebrate neurons [8], hippocampal slices [22], and cardiac cells [12]. The size of the neuronal cell body is around 10 μm, which is well-suited to standard photolithographic processes. Transparent conductive materials, such as indium-tin-oxide (ITO) have often been used to form electrode patterns on glass substrates [10]. This allows easy visual access to the target tissue and made simultaneous recording of electrical activity and intracellular calcium possible [13].

The basic fabrication process consists of three steps. First, electrode-array patterns are formed on insulating substrates. In the case of using ITO as the electrode material, features less than 10 μm wide can be easily formed by wet etching. Then, the insulating film layer is coated over the electrode patterns except at the recording sites. Finally, the surface of the recording sites are platinized to reduce interface impedance between solid and electrolyte solution. The electrical characteristics of this type of devices have been described by Bove et al [5]. When dissociated neurons are cultured on this substrate, poly-lysine and/or laminin is deposited on the surface to promote cell adhesion. Figure 1A shows an example of an electrode-array pattern, which contains 64 recording sites. A culture of cortical neurons on this substrate is shown in Fig. 1B.

Using planar electrode arrays, it is possible to carry out multi site recording of neural activity. Although another method, optical recording has much superior spatial resolution, with the present state of the art, the toxicity of the dyes to biological tissue and low S/N of the optical signals restrict the areas to which optical recording can be applied. The most important advantage of planar electrode array is its non-invasiveness and capability of multi site stimulation. Long term recording was carried out to see developmental changes of spontaneous activity in cortical neurons [15]. Multiple synaptic inputs to a single neuron were studied using multi site stimulation [14].

3 Modification of cortical network activity induced by tetanic stimulation

3.1 Spike trains recorded from dissociated neurons

As shown in Fig. 1B, dissociated cortical neurons extend lots of neurites and form densely connected networks. After about a single week *in vitro*, clear spike signals start to be recorded. It was reported that the bandwidth of these extracellularly recorded spike signals were between 100 Hz and 6 kHz [20]. Using present technology in electronic engineering, we can construct recording systems with sufficient time resolution. In our case, a 64 channel recording system with 50 kHz sampling rate for each was developed. An example of the recorded 64 traces is shown in Fig. 2A. A single pulse stimulus was applied from the site where a large artefact was shown and the evoked responses were recorded. Spikes of the amplitude between 50 and 300 μV were observed. The recorded spikes at each site were plotted on an amplitude-width plane [18]. This gave us clusters with clear peaks (Fig. 2B). At least three neurons were identifiable at this site. The total number of identified neurons was normally between 50 and 100.

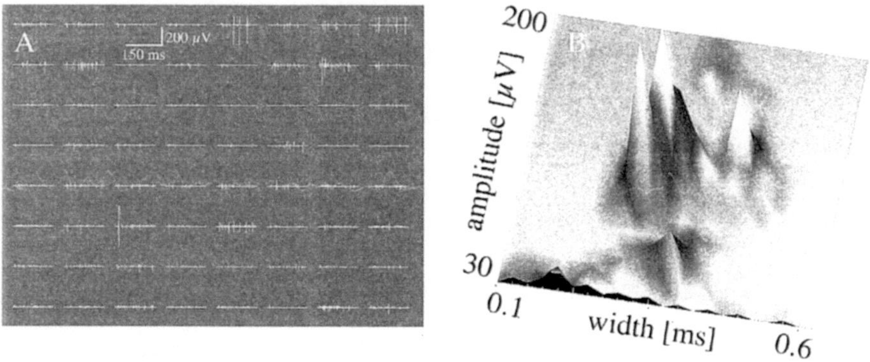

Figure 2 Multi site recording of neural spikes using planar electrode arrays. **A**. Evoked activity recorded at the 64 sites. **B**. Spike sorting in an amplitude-width plane. Spikes are plotted according to their amplitude and width. In this example, at least three individual neurons can be separated.

3.2 Effects of tetanic stimulation on the network activity

Modification of synaptic efficacy is widely believed to be a cellular basis for learning and memory. Both potentiation and depression have been reported and extensively studied particularly in the hippocampus [2, 6]. Most of these studies were focused on a change in the response of single neurons or local field potentials at single sites following tetanic stimulation, and little is known about network-wide changes in activity. The electrode-array method gives us a way to observe responses of a large number of cells simultaneously. Figure 3A shows a cultured cortical network formed on an electrode-array substrate. The substrate used for this had an about 1 x 3 mm recording area where 64 electrodes were located. A test stimulus was applied from a single site marked by * and the evoked responses were recorded. The numbers of detected spikes at each recording site were used as the measure for the response intensity and are displayed as brightness. The response before tetanus is shown in Fig. 3B. After tetanus, both the intensity of the response and the region where the response was observed increased (Fig. 3C).

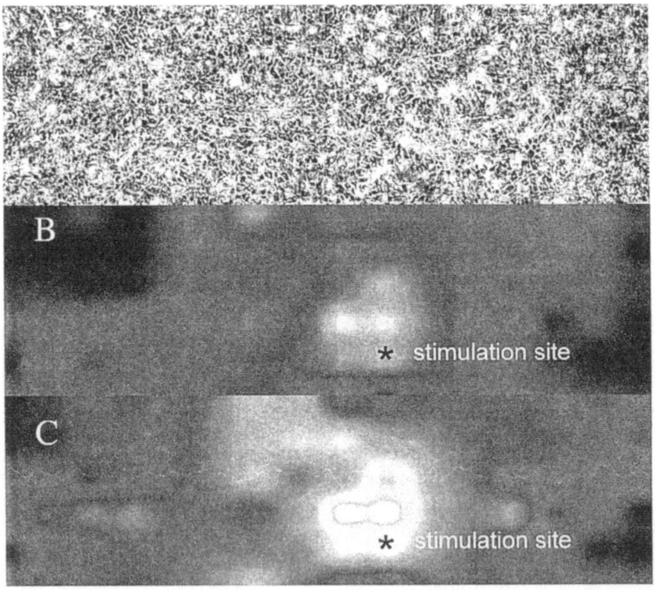

Figure 3 Effects of tetanic stimulation on the activity in dissociated cortical neurons. **A**. Cultured cortical networks formed on a electrode array substrate. **B**. A response evoked from a single site (marked by *) at the initial state. The number of detected spikes at each recording site was used as the measure for the response intensity and displayed as brightness. **C**. The same evoked response after tetanic stimulation. A clearly increased intensity of response was observed. The area where activity was evoked also extended.

4 Propagation of spontaneous activity in cortical slices

4.1 Field potentials recorded from slices

In slice preparations, intact structures are maintained two-dimensionally. Thus, when a slice is plated on a planar electrode array, we can expect to see the location where spontaneous activity initiates and how the spatial patterns of activity propagation corresponds to the structure of the tissue. Of course, the optical recording method has much superior spatial resolution and considerable temporal resolution. The advantage of the direct observation of electrical signals using electrode-arrays is its high S/N ratio. The intensity of the field potentials induced by neural activity in slices is as high as 1 mV. This enables us to record single spontaneous events without summation of repeated trials. It should be noted that the fact that the recorded electrical signals contain large components of field potentials, which is a characteristic feature in slice preparations, requires consideration of the volume conductor. Current source density analysis is often used for this purpose [11, 22].

4.2 Propagation of spontaneous activity in cortical slices

Cortical tissue has a layered structure composed of six layers. Connors reported in 1984 that bicuculline induced activity was initiated at IV/V layers and intrinsic bursting neurons in the layers were responsible for the activity [7]. More recently, however, optical recording has revealed that the cells in superficial layers play a key role for both initiation and propagation of the cortical activity [1, 21]. One reason for this discrepancy may be a technical limitation, the small number of recording sites in the former study and the summation of repeated trials in the latter. As mentioned above, the electrode-array method gave us a way of direct observation of a single spatially propagating spontaneous activity. Figure 4A shows a cortical slice plated on the substrate. This substrate has a recording area, the size of which is about 1 x 2 mm. Almost the whole area of the slice, which was taken from P2 Wistar rat pups, was covered by the 64 recording sites. White matter and pia sides are indicated in the figure. Kainic-acid application induced epileptiform activity, which was recorded by this configuration. In Fig. 4B, the time course of spatial distribution of electrical activity is shown. The intensity of the recorded signals is displayed as brightness. We see that the activity was first initiated near the surface and propagated along the superficial layer. Then it spreads over the slice including the deep layers. The triggering activity observed in the superficial layers was common to kainate or bicuculline induced activity, and to the spontaneous activity generated under culture conditions.

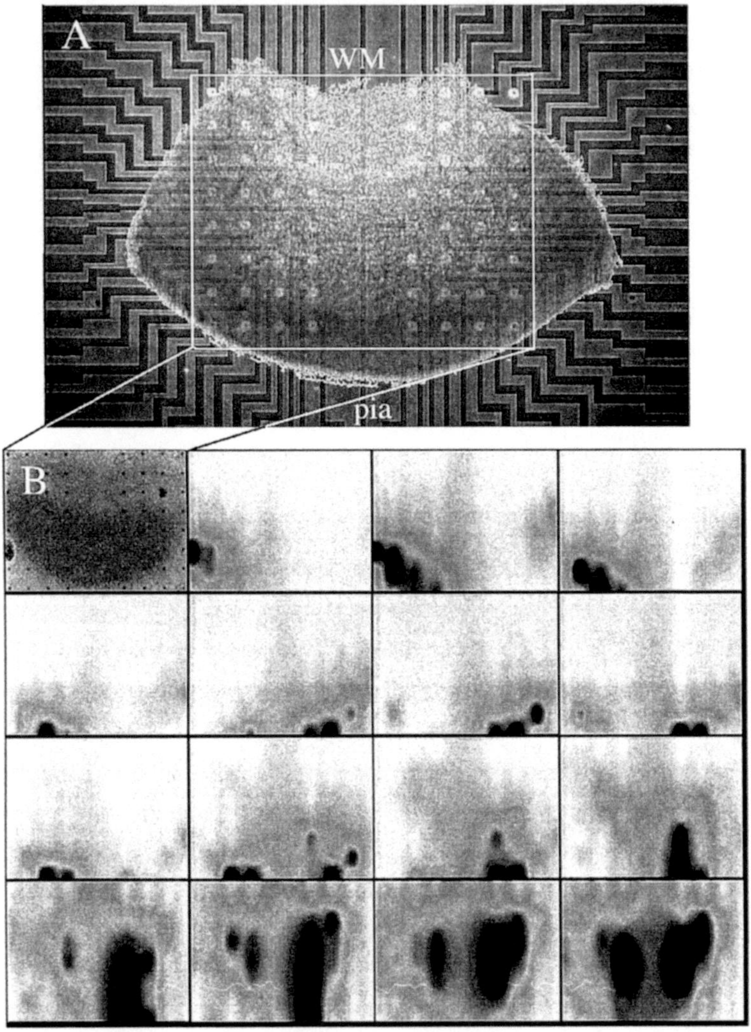

Figure 4 Recording spontaneous activity from a cultured cortical slice. **A**. A cortical slice plated on a planar electrode array. White matter and pia sides are indicated in the figure. Almost the whole area of the slice is covered by the 64 recording sites. **B**. Initiation and propagation of kainate-induced activity. The intensity of recorded field potentials were displayed as brightness. The time course goes from left to right, top to bottom. In this case, we can see that the activity was initiated near the surface and propagated along the superficial layer and then, spread almost all over the slice. The time of the total process was a few tens of ms.

5 Discussion

Twenty years have passed since planar electrode arrays were developed. The basic structure of the device itself has not been greatly modified since then. Major progress was made in the electronics for the signal recording and processing systems. High speed A/D converters as well as Digital Signal Processors opened up the possibility of real-time spike detection and sorting. This greatly reduced the amount of data and made it possible to analyze spike trains in groups of cells. Using this method, dynamics of systems composed of a large number of neurons will be analyzed.

Another point that should be noted is the combination of this method with other methods. Easy visual access to the tissue and non-invasiveness are the main advantage of the electrode arrays. Patch-clamp recording and/or fluorescence measurements including intracellular calcium concentration can be carried out together with extracellular activity recording. A combination of the high spatial resolution of optical recording with the fine temporal resolution of the electrode-array method will be a powerful, feasible approach for neurophysiological experiments.

With the present technical standard of micro fabrication technology, much finer structures than 10 μm wide electrodes can be made. Even three-dimensional structures are readily constructable using established fabrication techniques, such as anisotropic etching. So far, guidance of neurite outgrowth based on the difference of adhesiveness to the substrate surface has been achieved, forming geometrically controlled neural networks [16]. The next step will be establishment of methods for measuring physiological functions based on these individual techniques.

References

1. Albowitz, B., and Kuhnt, U.: Epileptiform activity in the Guinea-pig neocortical slice spreads preferentially along supragranular layers - recordings with voltage-sensitive dyes. Eur. J. Neurosci. **7**, 1273-1284 (1995)
2. Bashir, Z., Berretta, N., Bortolotto, Z., Clark, K., Davis, C., Freguelli, B. G., Harvey, J., Potier, B., and Collingridge, G. L.: NMDA receptors and long-term potentiation in the hippocampus. *In* The NMDA Receptor, Collingridge, G. L. and Watkins, J. C. eds. Oxford Univ. Press, 295-312 (1994)
3. BeMent, S. L., Wise, K. D., Anderson, D. J., Najafi, K., and Drake, K. L.: Solid-state electrodes for multichannel multiplexed intracortical neuronal recording. IEEE Trans. Biomed. Eng. **BME-33**, 230-241 (1986)

4. Bergveld, P.: Development of an ion-sensitive solid-state device for neurophysiological measurements. IEEE Trans. Biomed. Eng. **BME-17**, 70-71 (1970)
5. Bove, M., Grattarola, M., and Martinoia, S.: Coupling of networks of neurons to substrate planar microtransducers, a review. *In* Neurobiology, Torre V. and Conti, F. eds., Plenum, New York, 251-264 (1996)
6. Bliss, T. V. and Lømo, T.: Long-lasting potentiation of synaptic transmission in the dentate area of the anaesthetized rabbit following stimulation of the perforant path. J. Physiol. (Lond.) **232**, 331-356 (1973)
7. Connors, B. W.: Initiation of synchronized neuronal bursting in neocortex. Nature **310**, 685-687 (1984)
8. Curtis, A., Breckenridge, L., Connolly, P., Dow, J., Wilkinson, C., and Wilson, R.: Making real neural nets: design criteria. Med. Biol. Eng. Comput. **30**, CE33-CE36 (1992)
9. Gross, G. W., Rieske, E., Kreutzberg, G. W., and Meyer, A.: A new fixed-array multi-electrode system designed for long-term monitoring of extracellular single unit neuronal activity in vitro. Neurosci. Lett. **6**, 101-105 (1977)
10. Gross, G. W., Wen, W. Y., and Lin, J. W.: Transparent indium-tin oxide electrode patterns for extracellular, multisite recording in neuronal cultures. J. Neurosci. Methods **15**, 243-252 (1985)
11. Haberly, L. B. and Shepherd, G. M.: Current-density analysis of summed evoked potentials in opossum prepyriform cortex. J. Neurophysiol. **36**, 789-803 (1973)
12. Israel, D., Barry, W., Edell, D., and Mark, R.: An array of microelectrodes to stimulate and record from cardiac cells in culture. Am. J. Physiol. **247**, H669-H674 (1984)
13. Jimbo, Y, Robinson, H., and Kawana, A.: Simultaneous measurement of intracellular calcium and electrical activity from patterned neural networks in culture. IEEE Trans. Biomed. Eng. **BME- 40**, 804-810 (1993)
14. Jimbo, Y., Robinson, H., and Kawana, A.: Strengthening of synchronized activity by tetanic stimulation in cortical cultures: Application of planar electrode arrays. submitted to IEEE Trans.
15. Kamioka, H., Maeda, E., Jimbo, Y., Robinson, H., and Kawana, A.: Spontaneous periodic synchronized bursting during formation of mature patterns of connections in cortical cultures. Neurosci. Lett. **206**, 109-112 (1996)
16. Kleinfeld, D., Kahler, K., and Hockberger, P.: Controlled outgrowth of dissociated neurons on patterned substrates. J. Neurosci. **8**, 4098-4120 (1988)
17. Kovacs, G., Storment, C., Halks-Miller, M., Belczynski, C., Santina, C., Lewis, E., and Maluf, N.: Silicon-substrate microelectrode arrays for parallel recording of neural activity in peripheral and cranial nerves. IEEE Trans. Biomed. Eng. **BME-41**, 567-577 (1994)
18. Meister, M., Wong, R., Baylor, D., and Shatz, C.: Synchronous bursts of action potentials in ganglion cells of the developing mammalian retina. Science **17**, 939-943 (1991)
19. Najafi, K. and Wise, K. D.: An implantable multielectrode array with on-chip signal processing. IEEE J. Solid-State Circ., **SC-21**, 1035-1044 (1986)
20. Najafi, K.: Solid-state microsensors for cortical nerve recordings. IEEE EMBS Magazine **13**, 375-387 (1994)
21. Sutor, B., Hablitz, J., Rucker, F., and Bruggencate, G.: Spread of epileptiform activity in the immature rat neocortex studied with voltage-sensitive dyes and laser scanning microscopy. J. Neurophysiol. **72**, 1756-1768 (1994)
22. Wheeler, B. and Novak, J. L.: Current source density estimation using microelectrode array data from the hippocampal slice preparation. IEEE Trans. Biomed. Eng. **BME-33**, 1204-1212 (1986)

OPTICAL INSIGHTS INTO CEREBELLAR CIRCUITRY

Dana Cohen and Yosef Yarom
Department of Neurobiology, Life Science Institute and Center for Neural Computation, Hebrew University, Jerusalem, ISRAEL 91904

Introduction

It is commonly accepted that a deeper understanding of how neural networks function will depend on the ability to monitor brain activity with high temporal and spatial resolution. The need for such monitoring systems led, in the early seventies, to the development of optical imaging techniques using voltage-sensitive dyes (Ross *et al.*, 1974). The technique is based on the detection of light, emitted from special dye molecules that bind to membranes and fluoresce proportionally to the membrane potential. In its current state, the method is particularly suitable for monitoring synchronized activity in large neuron populations, as demonstrated here in the cerebellar cortex.

The cerebellar cortex is a modular lattice-like structure composed of basic units that repeat themselves throughout the cortex (Llinas, 1981; Ito, 1984). Several basic elements combine to form a functional unit, the size of which is determined by the spatial distribution of its inputs and outputs. Each unit consists of five cell types organized in three layers. The granule cells layer (the inner layer) receives an input from the mossy fibers and transmits an output to the molecular layer (the outer layer). The granule cells' axons bifurcate at different levels in the molecular layer to form the parallel fiber system. These parallel fibers synapse on the dendrites of Purkinje cells whose cell bodies form the middle layer of the cerebellar cortex. The axons of the Purkinje cells serve as the sole output of the cerebellar cortex. Three types of inhibitory interneurons modulate the activity in the cerebellar cortex; the basket and stellate cells lie in the molecular layer and the Golgi cells are situated in the granular layer. This morphological organization of the cerebellar cortex enables the activation of large populations of neurons in well-defined circuits and therefore can be readily studied using optical techniques.

Here we introduce a novel experimental system designed to explore the spatio-temporal organization of activity within the cerebellar cortex. We utilize optical imaging of a voltage-sensitive dye (RH-795) to monitor the evoked neuronal activity in an isolated cerebellar preparation. Optical signals are recorded from 128 photodiodes organized in a 12x12 array. Using a 10x or 40x objective, each element of the array detects light from an area of about 200μm x 200μm or 50μm x 50μm, respectively. The signals are amplified and digitized with 12-bit accuracy at maximal temporal resolution of 40 μsec (Microstar, DAP 3400a).

This report presents an analysis of the responses evoked by stimulating the input to the cerebellar cortex at the level of the parallel fibers. We first describe the response to surface stimulation which activates the parallel fibers, and then analyze the spatio-temporal organization of the cortical inhibitory system that these fibers activate.

A beam of parallel fibers.

The longitudinal morphology of the cerebellar cortex implies that stimulating the cerebellar surface will activate a beam of parallel fibers. Indeed, the classical work of Eccles and his colleagues (1967) demonstrated that such a stimulus elicits a beam of activity that propagates along the cerebellar folium at a constant velocity. The field potential of this activity is composed of action potentials in the parallel fibers as well as the post-synaptic responses of Purkinje cells and interneurons in the molecular layer.

The optical technique enables us to measure the activity simultaneously in several sites along the activated beam of parallel fibers. Figure 1 shows the responses at each recording site in a recording array. A stimulating electrode was placed on the surface of the cerebellar cortex to the right of the recorded area. The data are displayed as traces of change in fluorescence over time at each recording site. As expected, the stimulus generated a narrow beam of activity marked by the gray area. Depending on stimulus intensity, the length of this beam could reach 4 mm. The response, which first appeared on the right, closer to the stimulating electrode, was characterized by a fast positive wave (time to peak 2 msec, Fig. 2) followed by a slow decay (duration at half amplitude was 25-30 msec). The amplitude of the signal and the width of the beam increased with stimulus intensity (Fig. 3A). The amplitude of the initial peak decreased and its duration increased as the signal propagated along the beam (at a velocity of 0.2 m/s, Fig. 2).

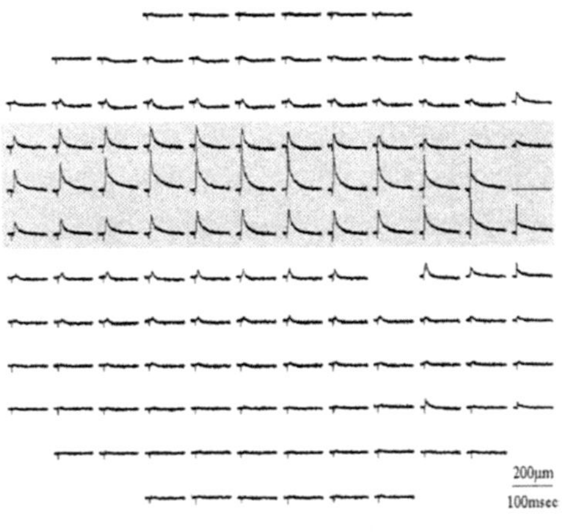

Fig. 1. Typical response to surface stimulation. Traces are arranged according to spatial location of recording site. Each trace shows changes in fluorescence as a function of time and is the average response of three identical stimuli repeated at a frequency of 0.2 Hz. A positive signal represents, as displayed, membrane depolarization. The stimulating electrode was located to the right of the recorded area. The gray area marks signals with large positive responses. The calibration bar gives the scale for both time and distance.

The field potential was recorded simultaneously with the optical measurements. The negative peak of the electrical responses occurred 0.5 msec before the peak of the optical responses (Fig. 3A). The optical responses, unlike electrical responses, most probably reflect mainly the post-synaptic activity induced by action potentials propagating in the parallel fibers. The post-synaptic nature of the optical signal was confirmed using specific synaptic blockers (Cohen, *et al.*, 1997). In these experiments, we were able to block completely the optical signals without interfering with the propagation of action potentials along the parallel fibers.

The absence of optical signals from the parallel fibers' action potentials is puzzling, especially since more than 40% of the membrane area in the molecular layer belongs to the parallel fibers. The reasons for such selective staining of the membranes by the voltage-sensitive dyes are yet to be investigated. It may be due to the specific properties of the different dyes or the different type of preparation - slices versus *in-vivo* or intact circuits. Ebner and Chen (1995) reported similar results using RH-414, 795 and RH-237. On the other hand, Konnerth *et al.* (1987) detected signals only from the parallel fibers when they used the voltage-sensitive dye, RH-482.

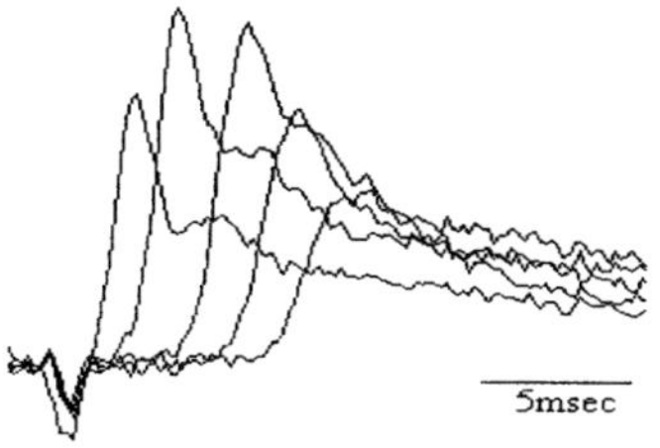

Fig. 2. Conduction velocity of the parallel fibers. Five traces located along the activated beam were superimposed. The distance between the first and the last diodes shown here was 2 mm.

Negative signals, reflecting inhibition, were usually observed on both sides of the beam (Fig. 1 and 3B). The amplitude of the lateral inhibition depended on stimulus intensity as shown in figure 3B. Occasionally, the lateral inhibition was preceded by a small positive signal. However, no correlation between the amplitude of the preceding excitatory peak and the amplitude of the inhibitory signal was found. It has been postulated (Eccles, *et al.*, 1967) that the function of this lateral inhibition, whose presence is to be expected from the structural organization of the stellate and basket cells, is to restrict the width of the beam. Our results clearly show a significant delay between the excitatory wave and the development of the lateral inhibition. Based on these results, we suggest that the role of the lateral inhibition is to limit the time interval in which interactions between adjacent cortical areas can occur.

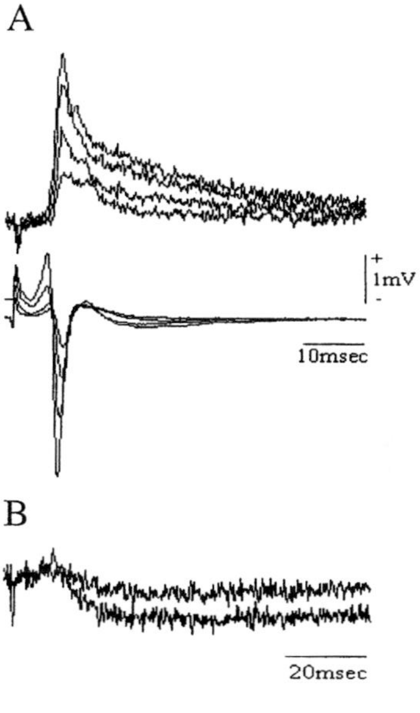

Fig. 3. Increase in stimulus intensity results in an increase in response amplitude. A: optical responses (upper traces) and electrical field potentials (lower traces) recorded at the same location at different stimulus intensities. B: lateral inhibition at two stimulus intensities.

On-beam inhibition

The molecular layer of the cerebellar cortex is interlaced with inhibitory interneurons. The excitatory input to these neurons is transmitted via the granular cells' ascending axon and the parallel fiber system. Their axons mostly terminate on Purkinje cells but mutual interactions between these neurons have also been reported (Palay and Chan-Palay, 1974). Little is known about the spatial and temporal organization of this network. One observation on the relationship between the activated beam of parallel fibers and the lateral inhibition was described above. We now describe the spatio-temporal organization of yet another type of inhibitory response. Since it is limited to the area covered by the beam of activated parallel fibers, we refer to it as the on-beam inhibition.

As described above, the response to a single stimulus of the cerebellar surface is characterized by a sharp positive peak followed by a prolonged decay. However, if a short train of such stimuli was applied, the positive response was usually followed by a prolonged negative response. As shown in figure 4, the prolonged negative signal occurred only at the center of the activated beam. Two different sets of experiments were performed to examine the nature of this response. First, we studied the effect of, a known GABAergic blocker, bicuculline, on the prolonged negativity. Bicuculline reversibly blocked this response. Second, we examined the effect of the prolonged negative signal on the response to a single test stimulus applied at different times. This test stimulus was delivered via a second electrode placed at a different location along the same beam of activity. The results in figure 5 show that the prolonged negativity inhibited the response to the single stimulus and that the time course of this effect was correlated with the time course of the prolonged negativity. Maximal inhibition (up to 70%) occurred at the peak of the negative response. Thus, we conclude that the prolonged negativity reflects an inhibitory response mediated via $GABA_A$ receptors.

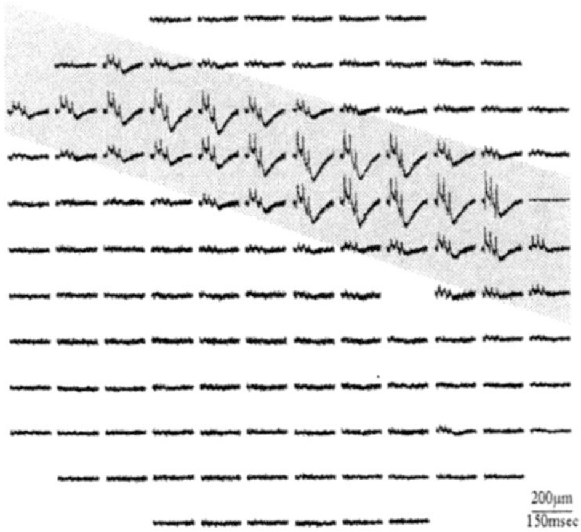

Fig. 4. On-beam inhibition elicited by a short train of surface stimulation. The stimulating electrode was located to the right of the recorded area. Each trace is an average response of 5 identical stimuli repeated at a frequency of 0.3 Hz. The stimuli consisted of three pulses at an interval of 20 msec. Data arranged as in Fig. 1.

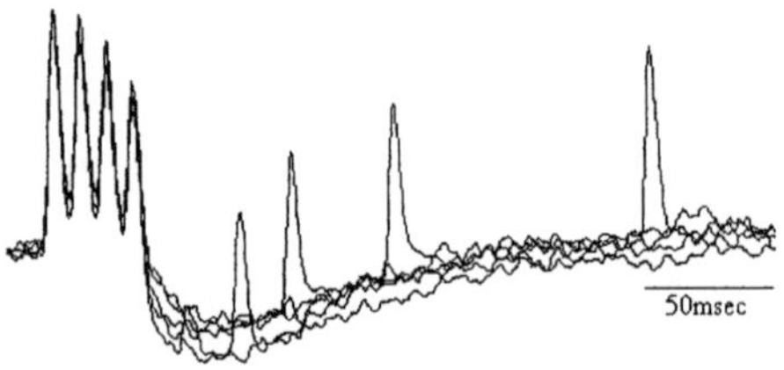

Fig. 5. The inhibitory effect of on-beam inhibition corresponds to the time course of the prolonged negativity (see text for details).

The amplitude of the on-beam inhibition depended on the stimulus intensity as shown in figure 6A. Increase in stimulus intensity increased both the peak of the positive response as well as the peak of the prolonged inhibition. Note however, that the duration of the signal and its time course of decay were unaffected. Quantitative analysis of the effect of the stimulus intensity is given in figure 6B in which the amplitude of the first positive response (circles) and the peak of the negative response (squares) are plotted as a function of the stimulus intensity. The similar dependence of the positive and negative responses on stimulus intensity resulted in a linear relationship between these two parameters (Fig. 6C, squares).

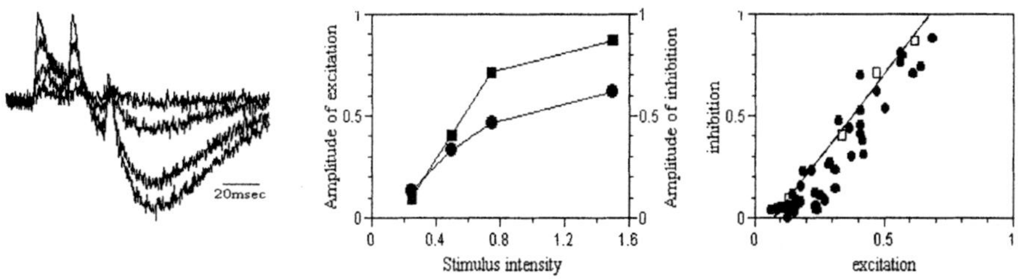

Fig. 6. Dependence of on-beam inhibition on stimulus intensity. A: responses to trains of three stimuli recorded at the center of the activated beam at four stimulus intensities. B: the normalized amplitude of the excitation (circles) and the inhibition (squares) as a function of the stimulus intensity. C: amplitude of the inhibition as a function of the excitation at the four stimulus intensities (squares). A similar linear relationship was observed between the excitation and inhibition in all active sites at one stimulus intensity (circles).

The same linear relationship was observed for the positive and negative responses recorded at different locations at one stimulus intensity (Fig. 6C, circles) and it holds for all stimulus intensities. The similarity of these curves demonstrates the power of the optical technique; it is now possible to replace a curve describing the intensity-dependence at a given location with a curve describing the same relationship but obtained simultaneously at several locations.

It is reasonable to assume either that both the excitatory and inhibitory signals are driven by the same source or that it is mainly the excitatory response that drives the inhibitory signals. Both possibilities fit the morphological organization of the cerebellar cortex. In the first case, the parallel fibers would be the common source of excitation to all types of neurons in the cerebellar cortex and, in the second case, the excitation of the inhibitory interneurons would be the source of the inhibitory effect. This latter possibility would mean that the positive responses are a mixture of excitation in Purkinje cells and the inhibitory interneurons, and the prolonged inhibition probably reflects the inhibition of Purkinje cells. Furthermore, the line fitted between the inhibition and the excitation intersects the excitation axis and this point of intersection may reflect the excitation of Purkinje cells. However, at present we cannot rule out the possibility that the prolonged inhibition is partially due to the inhibitory connection between the Golgi and granular cells.

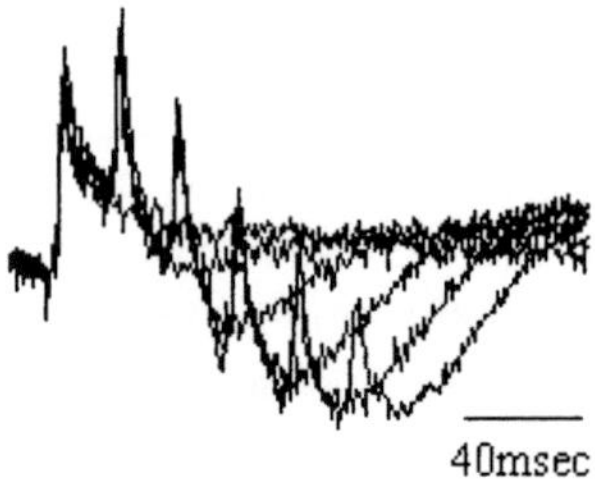

Fig. 7. On-beam inhibition increases as a function of the number of stimuli in the train. Six superimposed traces of different train duration are shown.

In order to analyze the kinetics of the on-beam inhibition, we studied the relationship between the number of stimuli in the train and the amplitude of the prolonged inhibition. Figure 7, which shows superimposed traces recorded at the center of the activated beam, demonstrates that increasing the number of stimuli increased the amplitude of the inhibition up to a saturating level of about 5 pulses. It also shows that a single stimulus is completely devoid of inhibitory signal. Thus, we conclude that the dependence of on-beam inhibition on the duration of the train reflects a rapid facilitatory mechanism. It is likely that this mechanism operates at the level of the inhibitory synapses between the inhibitory interneurons and the Purkinje cells.

Closing remarks

This short report on optical insights into cerebellar circuitry presents for the first time a powerful technique - the combination of an *in-vitro* preparation of an isolated cerebellum with optical measurements of voltage sensitive dyes. The isolated cerebellar preparation has outstanding advantages (Llinas, *et al.*, 1981): On one hand, like all *in-vitro* preparations, it allows experimental control, including physiological and mechanical stability as well as the accessibility to pharmacological manipulations. On the other hand, it maintains the integrity of the cerebellar circuitry. Thus, in a sense it bridges the gap between the commonly used slice preparations and the *in-vivo* approach to brain study. The advantages of optical measurements, which have been discussed in length in several reviews (Saltzberg, 1989; Grinvald, 1988; Wu and Cohen, 1993), are beyond the scope of this report. However, it is easy to grasp the possibilities that emerge in combining these two techniques, particularly in studying functional organization of neuronal circuitry.

Here we report on use of this technique to explore the organization of the cerebellar cortex. Armed with the ability to measure the activity over a relatively large area simultaneously, we are able to gain new insights into cerebellar circuitry. Optical measurements reveal that the lateral inhibition is delayed in time compared to the excitatory wave. Therefore, in contrast to the classical view, the lateral inhibition cannot determine the width of the activated beam of parallel fibers. We suggest instead that the lateral inhibition serves as a mechanism to shape the temporal interactions of adjacent cortical areas. That is, we hypothesize that it limits the time interval in which interactions can occur to a narrow window of 5 to 10 msec. The prolonged inhibition that ensues after this time interval will effectively prevent such interactions.

Analysis of the organization and dynamics of the on-beam inhibition allows further insight into cerebellar circuitry. This inhibition demonstrates a fast and efficient facilitatory mechanism that sets the temporal organization of the inhibitory system of the cerebellar cortex. The source of this signal is either the output of the inhibitory interneurons in the molecular layer or the output of the Golgi cells in the granule cells layer. In the first case, the inhibitory potential would occur mainly in Purkinje cells. It is also possible that a part is due to Purkinje cells axon collaterals or interconnections between inhibitory interneurons. The second possibility implies that the inhibitory potential arises in the granule cells. The effect of the on-beam inhibition on the test pulse (Fig. 5) strongly suggests that it arises in the molecular layer, the source of the excitatory response to the test pulse. However, it should be noted that a strong inhibition of the granular cell somata could theoretically interfere with conduction along the parallel fibers. However, regardless of the neuronal source of on-beam inhibition, it cuts the excitatory wave short. Facilitation ensures minimum interference of the inhibitory system with short duration excitatory activity while enabling a delayed and powerful inhibition of prolonged excitatory activity.

We propose that the inhibitory system of the cerebellar cortex is organized in spatial and temporal domains to enable neuronal interactions in a narrow window in time and space.

Acknowledgment: This study was supported by the US-Israel Binational Science Foundation and the Israel Academy of Sciences and Humanities.

References

Cohen, D., Segev, I. and Yarom, Y. 1997. Optical measurements reveal patchy organization of activity in the cerebellar cortex following mossy fiber stimulation. Soc. Neurosci. Abst., **23**, 1286.

Ebner, T. J. and Chen, G. 1995. Use of voltage-sensitive dyes and optical recordings in the central nervous system. *Progress in Neurobiology* 46: 463-506.

Eccles, J. C., Ito, M. and Szentagothai, J. 1967. *The Cerebellum as a Neuronal Machine*, (Springer-Verlag, Berlin).

Grinvald, A., Frostig, R. D., Lieke, E. E., and Hildesheim, R. 1988. Optical imaging of neuronal activity. *Physiological Review,* 1285-1366.

Ito, M. 1984. *The cerebellum and neural control.* Raven Press, New York.

Konnerth, A., Obaid, A. L., and Salzberg, B. M. 1987. Optical recording of electrical activity from parallel fibers and other cell types in skate cerebellar slices. *J. Physiol.* **393**, 681-702.

Llinas, R. 1981. Electrophysiology of the cerebellar networks. In *Handbook of Physiology-The Nervous System II.* V. B. Brooks ed.

Llinas, R., Yarom, Y., and Sugimori, R. 1981. Isolated mammalian brain *in-vitro*: New technique for analysis of electrical activity of neuronal circuit function. *Fed. Proc.* **40**, 2240-2245.

Palay, S. L. and Chan-Palay, V. 1974. *Cerebellar Cortex*, (Springer-Verlag, New York).

Ross, W. N., Salzberg, B. M., Cohen, L. B. and Davila, H. V. 1974. A large change in dye absorption during the action potential. *Biophys. J.* 14, 983-986.

Saltzberg, B. M. 1989. Optical recording of voltage changes in nerve terminals and in the fine neuronal processes. *Physiological Review* A. 507-526.

Wu, J. Y. and Cohen, L. B. 1993. *Fluorescent and Luminescent Probes for Biological Activity: A Practical Guide to Technology for Quantitative Real-Time Analysis.* W. T. Mason ed. (Academic Press, London).

Activity-dependent regulation of neuronal network excitability

G.J.A. Ramakers, F. Kloosterman, P. van Hulten, J. van Pelt and M.A. Corner

Neurons and Networks, Netherlands Institute for Brain Research, Graduate School Neurosciences Amsterdam, 1105 AZ Amsterdam, The Netherlands

Abstract. Electric activity plays a major role in the fine-tuning of neuronal connections during development. Since alterations in connectivity will in turn affect network activity, it is clear that neuronal network formation is the result of reciprocal interactions between the activity and the structure of the network. To investigate the role of electric activity in neuronal network development, we use primary cultures of dissociated fetal rat cerebral cortex. Following prolonged suppression of spontaneous electric activity in culture neuronal firing showed a strong increase in stereotyped burst firing, at the expense of variable non-burst firing. This mode of firing could be mimicked by blocking GABAergic inhibition, indicating that chronic suppression of electric activity induced a shift in the balance between excitation and inhibition, resulting in overexcitation. Chronic silencing induced a disproportionate decrease in GABA content, while the release of glutamate and aspartate was facilitated in early cultures. As a result, the ratio of stimulated release of excitatory versus inhibitory neurotransmitter was increased, in line with our hypothesis. Conversely, prolonged depolarization increased GABAergic staining intensity without affecting the density of GABAergic neurons. These data suggest that during brain development, compensatory mechanisms may operate, which serve to keep the level and/or pattern of electric activity within physiological limits. We propose a negative feed-back loop whereby electric activity stimulates the synthesis and release of BDNF, which through trkB receptors on GABAergic neurons stimulates GABAergic network activity, restraining overall network activity.

Keywords. neuronal network formation, electric activity, activity-dependent plasticity, GABA, neurotrophins, BDNF, TrkB

1 Introduction

One of the major questions in neurobiology is the nature of the mechanisms by which the enormously complex, yet precise, connectivity of the brain is established. And furthermore, how this connectivity is modified during learning processes, to enable the organism to adapt to changes in its environment. During development of the brain considerable remodelling takes place, which may serve to fine-tune neuronal connectivity to the functional demands by the environment. One of the major factors in the fine-tuning of neuronal connections is the electric activity of the neuronal network [14]. The fine-tuning of neuronal connectivity will in turn have physiological consequences. Thus, neuronal network formation is the result of reciprocal interactions between the activity and the structure of the network.

Our research is aimed at establishing the role of electric activity in neuronal network formation and the cellular mechanisms by which electric activity modulates neuronal connectivity. To investigate the role of electric activity in neuronal network development, we use primary cultures of dissociated fetal rat cerebral cortex. Such cultures have proved to be accessible model systems, which reliably mimic neuronal differentiation and network formation with a similar timing as in vivo [10]. Following trypsinization and mechanical dissociation of cerebral cortex neurons from embryonic day (E) 17 to 19, the neurons are plated onto coverslips or into tissue culture dishes coated with polyethylene imine or polylysine, and grown in a serum-free medium conditioned by astrocyte cultures [11]. Under these conditions, the neurons can be maintained for up to at least six weeks, while showing abundant synaptogenesis and spontaneous electric activity. Network development in these cultures is characterized by extensive neurite outgrowth during the first few weeks, followed by synapse formation from about the end of the first week onward [15]. Synaptic density peaks at 3 weeks in culture and then shows a partial decline which stabilizes after four weeks (Fig. 1).

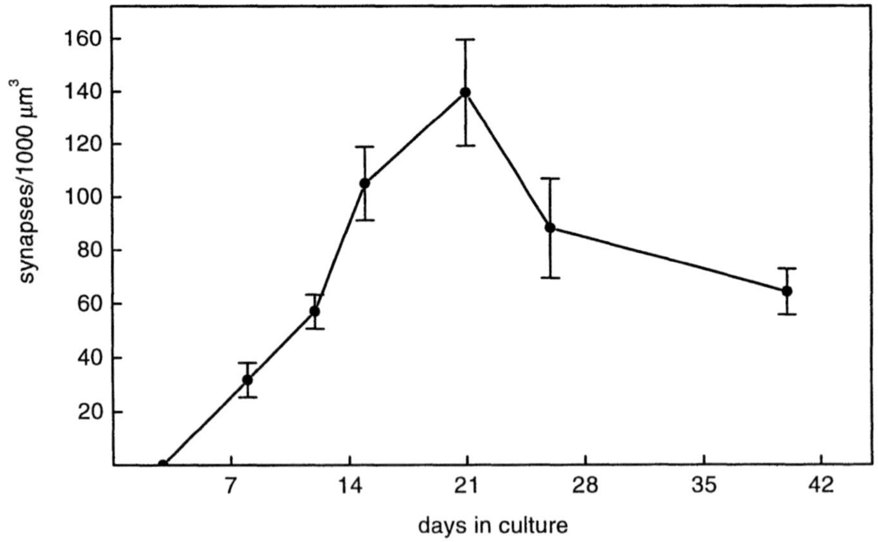

Fig. 1. Changes in synapse density in cultured cerebral cortex neurons as a function of culture age [15].

At the end of the first week neurons start to generate action potentials spontaneously (Fig. 2). The proportion of spontaneously active neurons and mean firing rates increase strongly during the second week in parallel with increasing synaptic densities. This spontaneous activity is a network phenomenon, as it is critically dependent on synaptic connectivity in culture. Blocking synaptic transmission by lowering calcium and increasing magnesium concentrations in the medium to 0.2 mM and 10 mM, respectively, almost completely suppresses electrical activity [9].

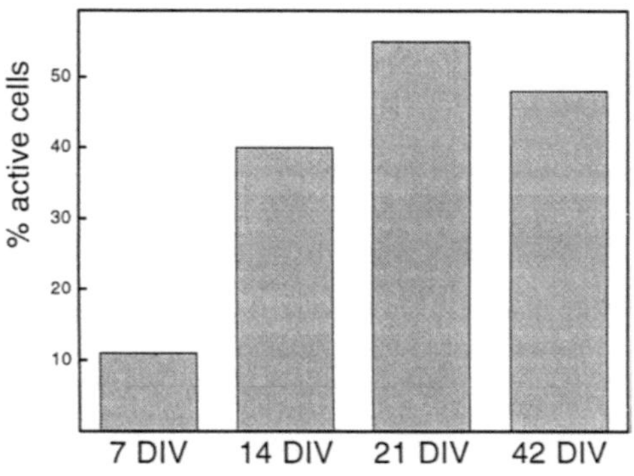

Fig. 2. Percentage of cells showing spontaneous action potential discharges as measured with extracellular electrodes.

2 Suppression of spontaneous electric activity induces hyperexcitability in culture

To investigate the importance of electric activity for structural and functional neuronal network development, spontaneous electric activity was chronically suppressed with the sodium channel blocker tetrodotoxin (TTX, 100 nM) from 5 or 6 DIV onward for up to several weeks. Washing out TTX resulted in an immediate return of spontaneous activity. Using extracellular electrodes, trains of action potentials were recorded from single neurons in previously silenced cultures and from control cultures, which had developed in the presence of electric activity [9]. Differences in firing patterns were quantitated using spike train analysis programs developed in our laboratory [4]. Following chronic silencing, the pattern of spontaneous electric activity had changed, with a strong increase in stereotyped burst firing, at the expense of variable non-burst firing (Fig. 3). This effect was seen at all time points investigated. Acute disinhibition of the cultures with picrotoxin (1 μM), an antagonist of the major inhibitory neurotransmitter GABA, induced similar stereotyped burst firing as seen after chronic silencing [9]. Thus, development in the absence of spontaneous electric activity apparently had resulted in a reduced inhibition within the network. Alternatively, since network activity is dependent on the balance between excitation and inhibition, an increase in excitatory network activity could not be ruled out. In addition to the increase in stereotyped burst firing, the proportion of active neurons as well as mean firing rates were increased, lending further support to the notion that the excitatory 'drive' of the network was increased.

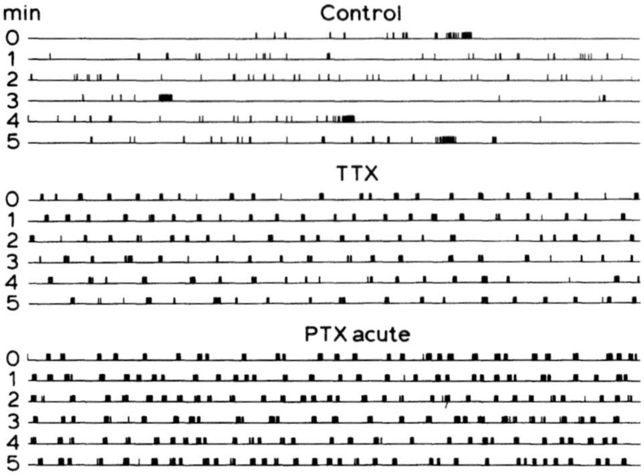

Fig. 3. Spike trains of spontaneously active single neurons recorded around the third week in culture for 6 min. following development in the presence of spontaneous activity (control), after chronic silencing (TTX, 100 nM), or in the acute presence of the GABA-A antagonist picrotoxin (PTX, 1 μM) [9].

The fact that the response of the neuronal network was in the opposite direction of the chronic treatment indicated that the response could be due to compensatory mechanisms. By increasing the ratio of excitation to inhibition, the network was trying to compensate for chronic silencing. When suppression was suddenly removed, the compensatory effects were unmasked. Possibly similar effects have been noted in spinal cord cultures [8]. When these cultures are chronically treated with glycine, which is the major inhibitory neurotransmitter in spinal cord, spontaneous activity also revealed a high incidence of stereotyped epileptiform burst firing after glycine was removed. Based on these observations we proposed that during neuronal network development, homeostatic mechanisms operate by which (spontaneous) electric activity regulates its own level within physiological limits, by modulating the balance between excitatory and inhibitory network activity [1].

3 Mechanisms underlying the activity-dependent regulation of the balance between excitation and inhibition

Both presynaptic and postsynaptic mechanisms could be responsible for changing the balance between excitation and inhibition in culture. Glutamate and aspartate are considered to be the major excitatory neurotransmitters in cerebral cortex. To investigate their role in spontaneous electric activity, glutamatergic neurotransmission through NMDA and non-NMDA receptors was specifically blocked with APV

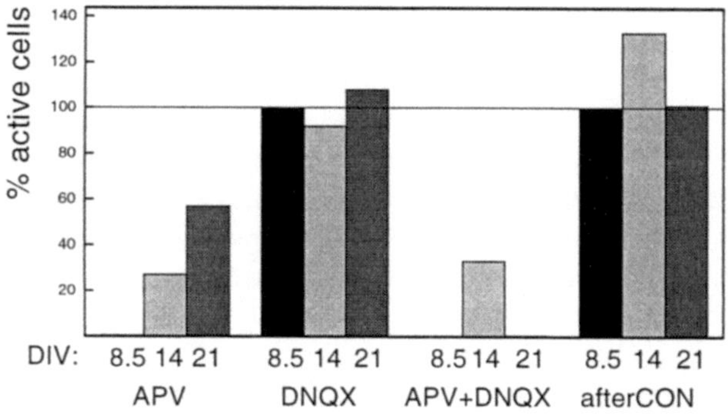

Fig. 4a. Percentage of spontaneously active cells, probed with extracellular electrodes in the presence of APV (100 μM), DNQX (100 μM) or both, or following washing out the antagonists. The precentage is expressed relative to the incidence of activity under control conditions (set at 100%) [12].

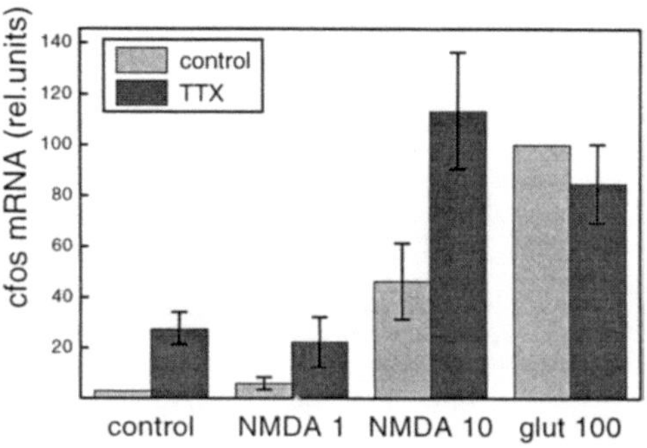

Fig. 4b. Induction of cfos mRNA by different concentrations of NMDA or glutamate following development of cerebral cortex neurons in the presence (control) or absence of spontaneous activity (TTX). Cfos mRNA is expressed as integrated optical densities after Northern blotting of equal amounts of total RNA; cultures are aged from 14 to 21 days.

(aminophosphonovalerate) and DNQX (dinitroquinoxaline), respectively [12]. Surprisingly, DNQX by itself did not reduce the incidence of spontaneous electric activity, although it did increase the incidence of burst firing (Fig. 4a). APV on the other hand completely blocked activity in early cultures. The suppressive effect of APV decreased gradually with time, and at three weeks activity was reduced by only 50%. The combination of APV and DNQX suppressed almost all activity between 7 and 21 DIV. Thus, spontaneous activity was initially completely dependent on NMDA-mediated synaptic transmission, and with time became less dependent on activation of NMDA receptors. Since the decline in the role of NMDA-mediated neurotransmission coincided with the onset and increase in spontaneous activity in culture, the possibility existed that electric activity reduced the role of NMDA receptors in spontaneous network activity. This possibility was tested by quantitating the induction of mRNA for cfos by glutamate (100 μM) or different concentrations of NMDA (ranging from 1 to 10 μM) in cerebral cortex cultures following development in the presence or absence of spontaneous electric activity (Fig. 4b). Basal cfos levels were increased after chronic TTX treatment, possibly due to the higher levels of spontaneous electric activity after the treatment. Whereas the response to 100 μM glutamate was the same for control cultures and silenced cultures, the cfos induction in response to 10 μM NMDA was greatly enhanced after TTX treatment. Thus, spontaneous electric activity is responsible for the specific downregulation of the role of NMDA receptors in mediating excitatory network activity.

In subsequent experiments the role of excitatory and inhibitory neurotransmitter release was investigated, by measuring the content and release of GABA, the main inhibitory neurotransmitter, and glutamate and aspartate, the main excitatory neurotransmitters in culture, relative to the neuronal marker protein NSE (neuron-specific enolase) [13]. GABA content and release showed a disproportionate decrease following chronic suppression of electric activity (Fig. 5), while the release of glutamate and aspartate was facilitated in early cultures (not shown). As a result, the ratio of stimulated release of excitatory versus inhibitory neurotransmitter was increased (Fig. 6), in line with our hypothesis.

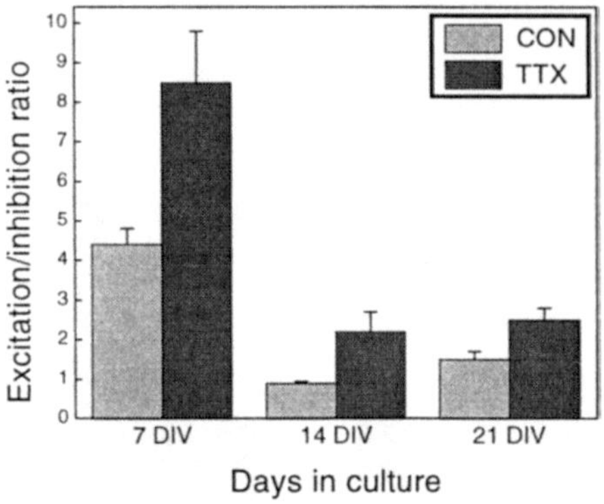

Fig. 6. Effects of chronic silencing of cerebral cortex cultures on the ratio of evoked release of glutamate and aspartate versus the evoked release of GABA. The excitation/inhibition ratio was defined as: (evoked release of glutamate + evoked release of aspartate) / evoked release of GABA. Release was evoked with 50 mM KCl for 3 min [13].

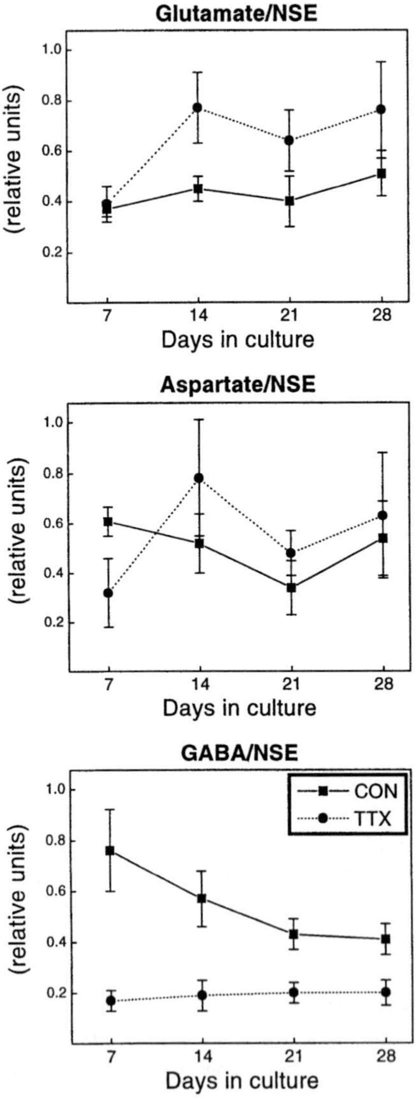

Fig. 5. Tissue content of aspartate (a), glutamate (b) and GABA (c), relative to the general neuronal marker NSE. Neurotransmitters were measured by HPLC, NSE was quantitated using Western blotting [13].

Since the GABAergic system appeared to be a major target of activity-dependent regulation, we investigated the effects of chronic depolarization on the development of GABAergic neurons at the structural level. Depolarization for three to four days stimulated the intensity of GABAergic staining without affecting the density of GABAergic neurons (Fig. 7). Preliminary evidence indicated an increase in the proportion of GABAergic synapses and in the size of GABAergic synapses after depolarization (Fig. 8). Furthermore, the intensity and number of neuropeptide Y stained neurons (a subpopulation of GABAergic neurons) was increased. These observations indicate that electric activity can regulate neuronal network activity through several distinct mechanisms, including downregulation of NMDA-mediated responses, modulation of transmitter release and stimulation of GABAergic network development.

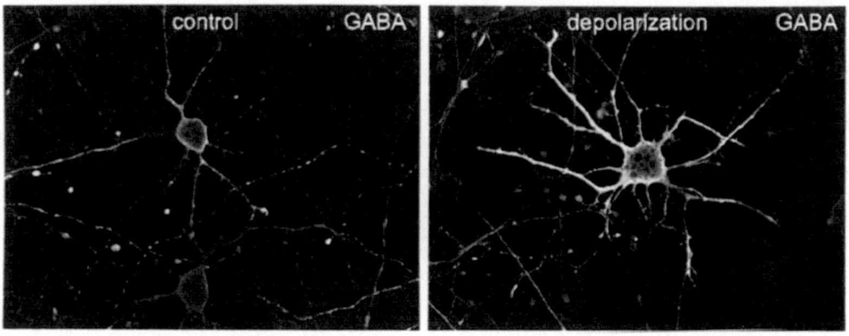

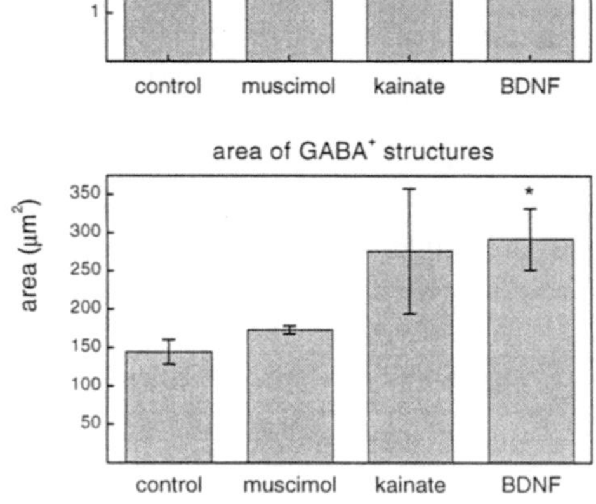

Fig. 7. Effects of depolarizing conditions on GABAergic network formation.

Above: depolarization with 25 mM KCl for 4 days increases the intensity of GABA-immunostaining. Both images were recorded using the same settings on the confocal microscope.

Left: muscimol (10 µM), kainate (10 µM) or BDNF (50 ng/ml) do not affect the proportion of GABAergic cells in culture. BDNF stimulates the morphological differentiation of GABAergic cells, as indicated by an increase in the area of GABA-immunoreactive structures. The increase by kainate is not significant due to the high standard error. Indicated are means +/- SEM; *: $p < 0.05$.

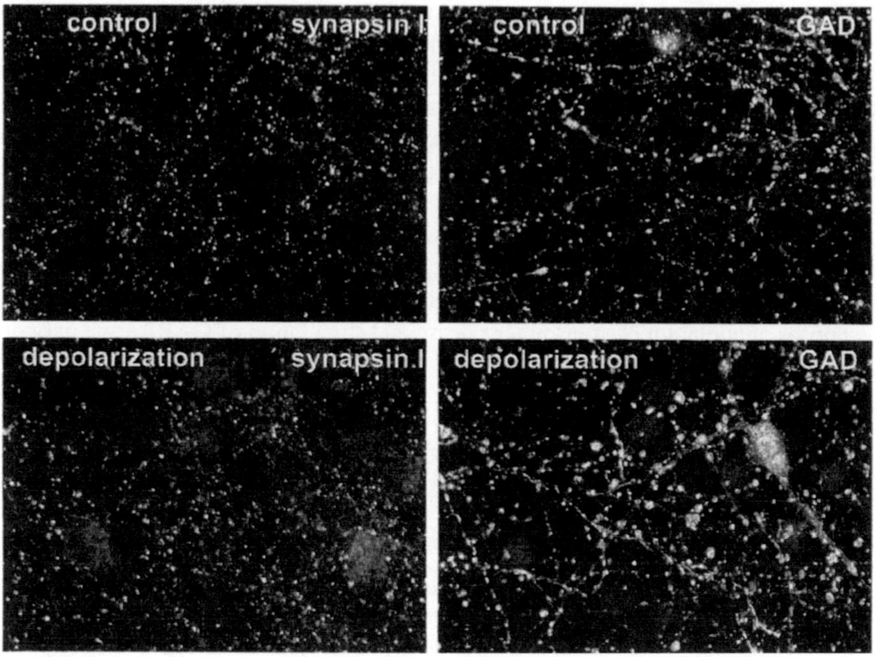

Fig. 8. Depolarization with 25 mM KCl specifically increases the size of GABAergic varicosities. However, the overall size of presumed presynaptic elements (synapsin I-positive varicosities) is not affected. The percentage of GAD (glutamic acid decarboxylase, the biosynthetic enzyme for GABA) positive varicosities which colocalize with synapsin I (i.e., the GABAergic synapses) increases as a result of depolarization.

5 Involvement of BDNF/TrkB in activity-dependent regulation of neuronal network excitability

Electric activity has been shown to stimulate the synthesis of certain neurotrophins, in particular brain-derived neurotrophic factor (BDNF)[5]. Evidence indicates that depolarization increases BDNF release from hippocampal and cerebral cortex neurons in tissue culture [2, 3]. In turn, BDNF was shown to promote the differentiation of GABAergic neurons with at least some degree of selectivity [7]. We observed a similar stimulation of GABAergic differentiation in cerebral cortex cultures (Fig. 7, lower part). These observations suggest a negative feed-back loop, whereby electric activity stimulates the synthesis and release of BDNF, which through trkB receptors on GABAergic neurons stimulates GABAergic network activity, restraining overall network activity (Fig. 10).

To investigate the role of TrkB in activity-dependent control of network excitability, we examined the effects of prolonged depolarization on the expression

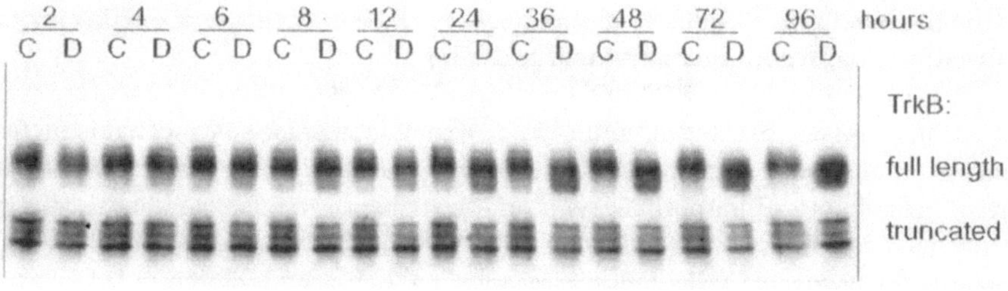

Fig. 9. Expression of full length and truncated TrkB in two week old cerebral cortex cultures upon depolarization with 25 mK KCl for 2 up to 96 hours. The 'smearing' of the TrkB bands is probably due to different degrees of phosphorylation or glycosylation.

of TrkB in cerebral cortex cultures. Previous work had shown that epileptic seizures stimulate trkB expression in vivo [6]. Western blotting showed that depolarization with 25 mM potassium chloride for 24 hours or longer stimulated the expression of full length TrkB, but not of the truncated receptor (Fig 9). Depolarization also decreased the apparent molecular weight of full length trkB, indicating a reduction of receptor phosphorylation or glycosylation. It will be important to establish how these changes affect signaling through TrkB, since concurrent stimulation of BDNF and TrkB expression by electric activity could have synergistic effects on TrkB-mediated signal transduction.

Since BDNF has also been shown to stimulate excitatory neuronal differentiation and neurotransmission, it remains to be established if and how neurotrophin signaling could differentially affect excitatory and inhibitory network formation. One possibility could be a higher expression of trkB, the receptor for BDNF, in GABAergic neurons compared to excitatory neurons. Alternatively, TrkB-mediated signal transduction could be more efficient in GABAergic neurons, or BDNF might be preferentially secreted by nerve terminals ending on GABAergic neurons.

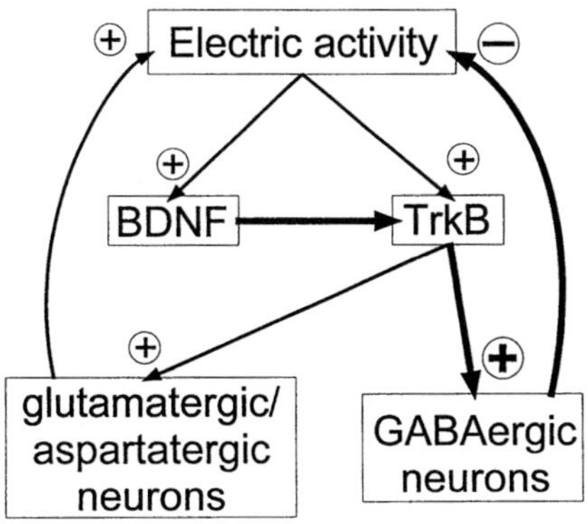

Fig. 10. Schematic outline of our working hypothesis of the activity-dependent regulation of neuronal network excitability through modulation of GABAergic network development.

6 Conclusions: role of activity-dependent regulation of network excitability in functional neuronal plasticity and pathology

Our results suggest that during brain development compensatory mechanisms may operate, which serve to keep the level and/or pattern of electric activity within physiological limits. Control of the levels of (spontaneous) activity within the cerebral cortex may be important in information processing and mechanisms of synaptic plasticity. Disturbance of the balance between excitation and inhibition is known to give rise to epileptic seizures. In most forms of epilepsy in humans the cause of the susceptibility to develop seizures is not known. The homeostatic mechanisms proposed here provide a framework to investigate a possible involvement of BDNF/TrkB in the etiology of epilepsy.

Although the data are far from complete, they suggest an activity-dependent regulation of neuronal network excitability, involving neurotrophin signaling and differential stimulation of excitatory and inhibitory neuronal network differentiation. In addition to the widely studied effects of activity on the potentiation of excitatory neurotransmission, regulation of inhibitory neuronal network development and function may play a major role in neuronal information processing and learning mechanisms.

References

1. Corner, M.A., Ramakers, G.J.A.: Spontaneous firing as an epigenetic factor in brain development - Physiological consequences of chronic tetrodotoxin and picrotoxin exposure on cultured rat neocortex neurons. *Dev. Brain Res.* 65, 57-64 (1992)
2. Ghosh, A., Carnahan, J., Greenberg, M.E.: Requirement for BDNF in activity-dependent survival of cortical neurons. Science **263**, 1618-1623 (1994)
3. Goodman, L.J. et al.: Regulated release and polarized localization of brain-derived neurotrophic factor in hippocampal neurons. Mol. Cell. Neurosci. **7**, 222-238.
4. Habets, A.M.M.C. et al.: Spontaneous neuronal firing patterns in fetal rat cortical networks during development in vitro: a quantitative analysis. Exp. Brain Res. **69**, 43-52 (1987)
5. Lindholm, D. et al.: Activity-dependent and hormonal regulation of neurotrophin mRNA levels in the brain--implications for neuronal plasticity. J. Neurobiol. **25**, 1362-1372 (1994)
6. Merlio, J.P. et al.: Increased production of the TrkB protein tyrosine kinase receptor after brain insults. Neuron **10**, 151-164 (1993)
7. Mizuno, K., Carnahan, J. and Nawa, H.: Brain-derived neurotrophic factor promotes differentiation of striatal GABAergic neurons. Dev. Biol. **165**, 243-256 (1994)
8. Nelson, P.G. et al.: Mouse spinal cord in cell culture: IV. Modulation of inhibitory synaptic function. J. Neurophysiol. **40**, 1178-1187 (1977)

9. Ramakers, G.J.A., Corner, M.A., Habets, A.M.M.C.: Development in the absence of spontaneous bioelectric activity results in increased stereotyped burst firing in cultures of dissociated cerebral cortex. Exp. Brain Res. **79**, 157-166 (1990)
10. Ramakers, G.J.A.: Development of fetal rat cerebral cortex in tissue culture and the role of bioelectric activity. University of Amsterdam, The Netherlands, Ph.D. thesis 1991
11. Ramakers, G.J.A. et al.: Development of neurons and glial cells in cerebral cortex cultured in the presence or absence of bioelectric activity: I. Morphological observations. Eur. J. Neurosci. **3**, 140-153 (1991)
12. Ramakers, G.J.A. et al.: A developmental decrease in NMDA-mediated spontaneous firing in cultured rat cerebral cortex. Int. J. Dev. Neurosci. **11**, 25-32 (1993)
13. Ramakers, G.J.A.et al.: Activity-dependent plasticity of inhibitory and excitatory amino acid transmitter systems in cultured rat cerebral cortex. Int. J. Dev. Neurosci. **12**, 611-621 (1994)
14. Shatz, C.J.: Impulse activity and the patterning of connections during CNS development. Neuron **5**, 745-756 (1990)
15. Van Huizen, F., Romijn, H.J., Habets, A.M.M.C.: Synaptogenesis in rat cerebral cortex cultures is affected by during chronic blockade of spontaneous bioelectric activity by tetrodotoxin. Dev. Brain Res. **19**, 67-80 (1985)

Spatial and Temporal Regulation of BDNF Action in Synaptic Plasticity

Enrico Tongiorgi[1,2], Massimo Righi[1], Mauro Benedetti[1] and Antonino Cattaneo[1]

[1]International School for Advanced Studies (SISSA), 34014 Trieste, Italy
[2]Department of Biology, University of Trieste, 34100 Trieste, Italy

Abstract

Molecules involved in synaptic plasticity need to be targeted to the appropriate subcellular domains and to be expressed at the right time during modulation of the synaptic activity. We show that the mRNAs for BDNF and TrkB have a dendritic localisation in cultured hippocampal neurons. Exposure of cells to high potassium for different times, differentially increases both the dendritic extension of BDNF and TrkB mRNAs and their protein levels in the distal portion of dendrites. The activity dependent modulation of mRNA targeting and protein accumulation in the dendrites may provide a mechanism for achieving local regulation of the activity of neurotrophins and their receptors with a time course compatible with their involvement in synaptic potentiation. Further, we put the hypothesis forward that the enhanced mRNA dendritic content for BDNF and TrkB might be involved in the synaptic tagging of a subset of dendritic branches.

Keywords: neurotrophins, dendritic mRNA, hippocampal neurons, TrkB, BDNF

1 Introduction

Synapses located on a single neuron can undergo long-lasting changes in synaptic transmission having very different information content. In order to maintain this information spatially and temporally distinct, neurons must be able to process selectively at each synapse, or group of synapses, different events such as potentiation or depression. Long-term synaptic modifications require a timed synthesis of new mRNAs and proteins and their targeted transport to the appropriate sites. The neurotrophin BDNF (Leibrock et al., 1989) is a member of the nerve growth factor family including also NGF, NT3 and NT-4/5, (reviewed in Levi-Montalcini, 1987, Barde, 1990; Davies, 1994; Lewin and Barde, 1996) and has been recently shown to be one of those molecules required for synaptic plasticity (reviewed in Thoenen, 1995; Bonhoeffer, 1996). In particular, recent evidence shows that hippocampal LTP in the CA1 region of BDNF-knock out mice is severely impaired (Korte et al., 1995). Indeed, BDNF can have direct effects on synaptic transmission (Lohof et al., 1993; Knipper et al., 1994a; 1994b; Leßmannn et al., 1994; Levine et al., 1995; Carmignoto et al., 1997) and a recent study has shown that BDNF (as well as NT-3) is able to produce a long-lasting enhancement of synaptic transmission in the rat hippocampus (Kang and Schuman, 1995; Kang and Schuman, 1996). BNDF mRNA expression itself is regulated by the glutamatergic, GABAergic and cholinergic neurotransmitter systems (Zafra et al., 1990; 1991; 1992). Moreover, the expression of BDNF mRNA is upregulated by physiological stimuli (Castrén et al., 1992) as well as by stimuli that lead to hippocampal long term potentiation (LTP) (Patterson et al., 1992; Castrén et al., 1993; Dragunow et al., 1993; reviewed in Lindholm et al., 1994).

An important first consequence of this role of neurotrophins in synaptic plasticity, is that the expression and the release of neurotrophins, as well as the expression or the responsiveness of receptors, should be specifically localised to the more active subcellular domains of neurons. The second consequence is that availability of these molecules at activated synapses should occur within a time scale consistent with the variation in the synaptic plasticity observed. Consistently with the first of these predictions, it has been recently demonstrated that NGF and BDNF are stored in the somatodendritic compartment of hippocampal neurons (Dugich-Djordjevic et al., 1995; Wetmore et al., 1994), and that their secretion from dendrites occurs mainly through an activity dependent, regulated pathway (Blöchl and Thoenen, 1996; Goodman et al., 1996), unlike the constitutive secretion of NGF, which occur mainly at the soma and the proximal part of the dendrites (Blöchl and Thoenen, 1995, 1996).

The sorting of specific mRNAs to translationally competent dendrites (reviewed in: Steward; 1994, 1997, Johnston, 1995) is an important way to localise the synthesis of proteins close to their sites of function. An increasing number of dendritically targeted mRNAs have been recently identified including the mRNAs encoding for MAP2, arc, CaMKinaseII α-subunit, IP3-receptor, NMDAR1 subunit, Glycin receptor α-subunit, vasopressin and dendrin (Garner et al., 1988; Lyford et al., 1995; Burgin et al., 1990; Furuichi et al., 1993; Gazzaley et al., 1997; Racca et al., 1997; Prakash et al., 1997; Herb et al., 1997). Moreover, several recent studies demonstrate the existence of a dendritic synthetic machinery for protein synthesis including ribosomes, tRNA and translation factors (Steward and Levy, 1982; Tiedge and Brosius, 1996), endoplasmic reticulum and Golgi-like apparatus (Spacek and Harris, 1997; Torre and Steward, 1996). A direct and unequivocal demonstration of local protein synthesis in fully developed dendrites physically isolated from the cell body, has been provided by Torre and Steward (1992) and more recently, the local translation of a reporter mRNA transfected into isolated growth cones was demonstrated (Crino and Eberwine, 1996).

We have recently demonstrated by means of in situ hybridisation that the mRNAs for BDNF and for TrkB belong to the group of dendritic mRNAs (Tongiorgi et al., 1996). Moreover, we have demonstrated that in rat hippocampal neurons in culture the mRNA targeting for both BDNF and TrkB, is regulated by electrical activity and contributes at least in part to the local synthesis of the corresponding proteins (Tongiorgi et al., 1997). In the present work we have extended these previous studies, by analysing more in detail the time dependence of BDNF and TrkB mRNA targeting within the dendrites. The time course and the localization of the events observed in the present study and in our previous work are discussed in the context of the spatial and temporal regulation of synaptic plasticity by neurotrophins in general, and by BDNF in particular.

2 Results and conclusions

The subcellular localization of the mRNAs encoding for BDNF and its receptor TrkB was determined by in situ hybridisation in rat hippocampal neurons maintained for one week in culture. In control conditions the mRNAs for both proteins are localised in the somatodendritic compartment (Fig. 1 A) and the maximal distance of the dendritic labelling (MDDL) measured on the entire population of dendrites is in average 32.5 µm for BDNF and 31.6 µm for TrkB mRNA (Table I), in accordance with our previously published data (Tongiorgi et al., 1997).

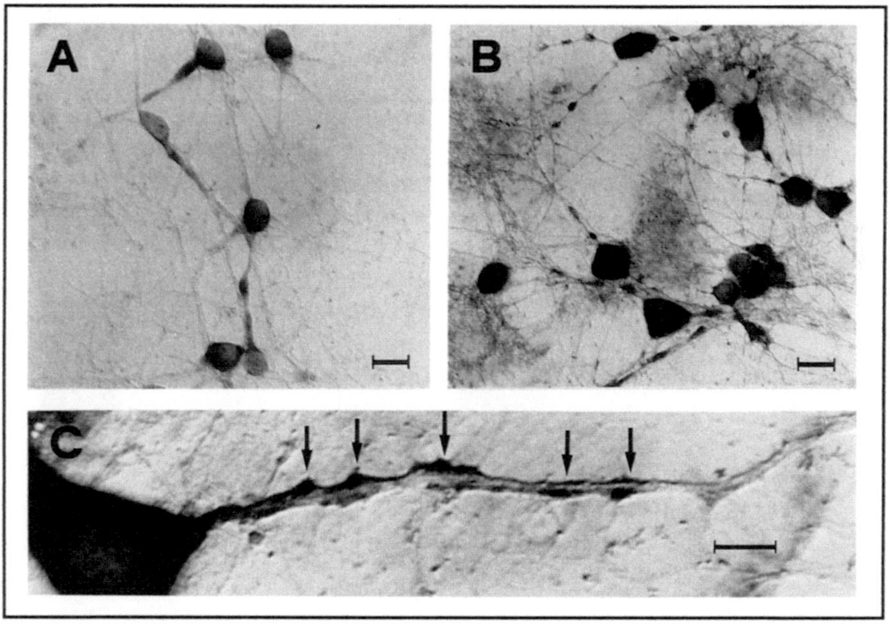

Fig. 1. Non-radioactive in situ hybridisation on rat hippocampal neurons cultured for 7 d. A) BDNF mRNA is localised in the soma and in the proximal portion of the dendrites in control conditions. B) After 3 h in 10 mM KCl BDNF mRNA is localised in the soma and extends into the distal portion of the dendrites. C) BDNF mRNA is accumulated in the dendritic varicosities (arrows). Scale bar is 10 µm in C and 20 µm in A,B.

These values demonstrate that the localisation of both mRNAs is dendritic, in agreement with a previous work (Martone et al., 1996) showing that the mRNA encoding for the Calmodulin Kinase II-α subunit (CaMKII-α), a typical dendritic mRNA, reaches the mean distance along the dendrites of 40 μm from the cell body, while the CaMKII-β subunit mRNA, which is not dendritic, reaches the average distance of 15 μm.

We have previously demonstrated that the depolarisation of hippocampal neurons, obtained by increasing the extracellular KCl concentration from 3.5 mM to 10 mM, is able to increase the MDDL for both mRNAs (Tongiorgi et al., 1997). To investigate the effects of the different times of depolarisation on the BDNF and TrkB mRNAs localisation, hippocampal neurons in culture were incubated with 10 mM for 3 hours continuously or for only 10 minutes followed by an incubation in control medium for 2 hours and 50 minutes. In this way, the total incubation time in both experimental conditions was the same. This would allow the dendritic mRNAs to be transported within the cells for the same time. After a depolarisation with 10 mM KCl for 10 minutes or for 3 hours a strong increase in the extent of dendritic staining for both BDNF and TrkB mRNAs can be observed (Fig.1 B, C).

The maximum distance from the cell body at which labelled BDNF and TrkB mRNA could be detected in individual dendrites was determined. For these measurements, cells with a well identified dendritic tree were randomly chosen within the cell population; results are shown in Table I and Fig. 2.

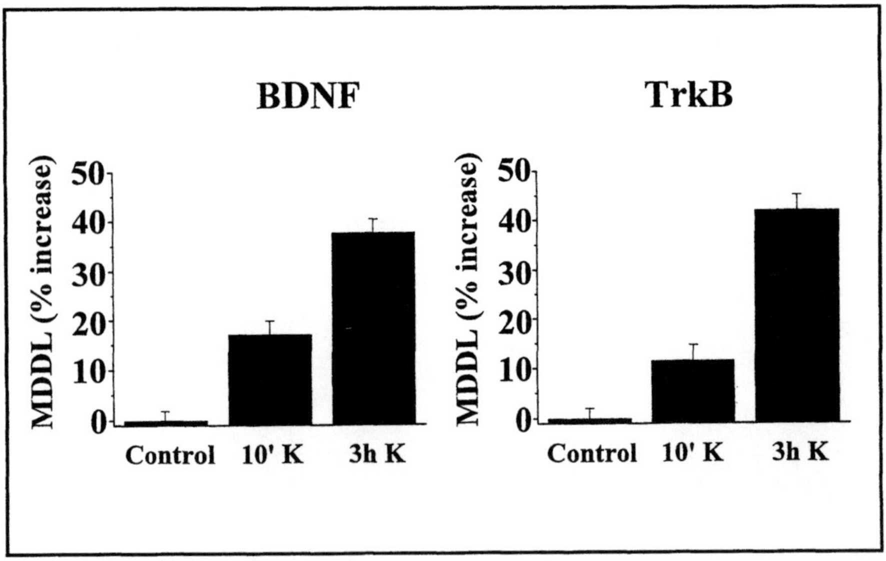

Fig. 2. Quantification of the MDDL at various times of incubation in 10 mM KCl. Values are expressed in percentage of the control values. For both BDNF and TrkB there is a progressive MDDL increase from 10 minutes depolarisation (10'K) to 3 hours of continuous depolarisation (3hK).

The increase in MDDL is much greater after 3 hours of continuous incubation with 10 mM KCl reaching similar average distances for both BDNF (44.8 μm) and TrkB (44.9 μm), corresponding to a 38% and a 42% increase, respectively with respect to the controls (Table I, Fig. 2). Thus, the subcellular localisation of BDNF and TrkB mRNAs results to be differentially regulated by different duration of depolarisation.

Table I

BDNF	C	32.5 μm	N= 577	SE± 0.59	-
	10'K	38.1 μm	N= 296	SE±0.90	$p<0.001$
	3hK	44.8 μm	N= 345	SE± 0.88	$p<0.001$
TrkB	C	31.6 μm	N= 670	SE± 0.68	-
	10'K	35.2 μm	N= 300	SE± 0.99	$p<0.01$
	3hK	44.9 μm	N=360	SE± 0.95	$p<0.001$

It has recently been shown that BDNF is able to produce a long-lasting enhancement of synaptic transmission in the hippocampus that, unlike other forms of plasticity, exhibited an immediate requirement for dendritic local protein synthesis (Kang and Schuman, 1996, Schuman, 1997). It is conceivable, that the presence of BNDF and TrkB mRNAs in the dendritic compartment might contribute to the local synthesis of these two proteins in response to enhanced synaptic activity and/or increased release of BDNF itself.

In a recent study, Frey and Morris (1997) have demonstrated that synapses of hippocampal neurons potentiated by a tetanus stimulation become tagged or marked and that this event is required for the formation of a long lasting potentiation. Subsequent to this early tagging event, to achieve the phase of consolidation of long term potentiation (late-LTP), there is a need of new protein synthesis otherwise the initial potentiation declines within one hour from the tetanus stimulation. The synaptic tag forms in the absence of protein synthesis and lasts about 2-3 hours in the experiments of Frey and Morris. Interestingly, it has been shown that the inhibition of protein synthesis does not prevent the dendritic transport of mRNA for several hours (Kleiman et al., 1993) and, in line with these results, we have previously shown that also an mRNA synthesis blockade for 3 hours does not prevent the activity dependent transport of BDNF and TrkB mRNAs into the dendrites. Thus, there is the possibility that the protein synthesis independent tag might be represented, at least in part, by the transport of specific mRNAs along the dendrites. In line with this, our data shows that 10 minutes stimulation can induce a long lasting increased mRNA transport for BDNF and TrkB. It has to be noted that when the hippocampal cultures were analysed after a 30 minutes total incubation time in 10 mM KCl we could also observe an increase in the dendritic localisation of the TrkB mRNA and to a lesser extent of BDNF mRNA (Tongiorgi et al., 1997). The effect of KCl is largely mediated by the neurotransmitter glutamate (Tongiorgi et al., 1997) which is known to be massively released upon KCl depolarisation. On the other hand, upon depolarisation, not only glutamate but also the neurotrophins might be rapidly released from the dendrites (Blöchl and Thoenen, 1995) contributing therefore to the synaptic signalling involved the potentiation of the synaptic connections. Indeed, a recent study has shown that small granules of RNA enter into the dendrites within 10 minutes from the beginning of the incubation with the neurotrophin NT3 (Knowles et al., 1997). In addition, we have preliminary data consistent with the idea that also BDNF is able to modulate the local dendritic availability of its own mRNA and that of its own receptor TrkB (Righi et al., 1998). It is worth noting that in the experiments by Knowles et al. (1997), the quantities of RNA (a few granules) that are displaced from the cell body or the proximal dendrites to the more distal regions of the dendrites are small but this might have an enormous impact on the composition of the pool of proteins located at the synapses. On this regard, it is remarkable that several dendritic mRNAs encode for proteins such as for example the NMDA receptor, and TrkB that have been found in the postsynaptic densities (Wu et al., 1996). From these studies, the emerging picture is one in which the time scale of the reorganisation of the subcellular localisation of the mRNAs within the hippocampal neurons, would be well consistent with the model of synaptic tagging proposed by Frey and Morris.

From the available data it is not possible to determine whether the activity dependent transport of mRNAs in the dendrites might be specifically directed towards the more active synapses. A high specificity of targeting at the level of a single synapse seems however to be unlikely since Frey and Morris (1997) demonstrated the presence of a synaptic cross talk due to the sharing of the tagging signal between neighbouring synapses. In addition, in a recent paper, it has been shown that synapses in close proximity to a site of stimulation are also potentiated regardless of their own history of activation, showing that there is no input specificity at a distance of less than 70 μm from the potentiated synapse (Engert and Bonhoeffer, 1997). Hence, a very sharp spatial restriction of the mRNA targeting or local protein synthesis within the dendrites is unlikely. It is conceivable, therefore, that the spreading of synapse potentiation might be due to an activation signal (possibly Ca^{2+}) that would enhance both mRNA targeting and local protein synthesis in a subset of dendritic branches, but not in a single one.

As a first step towards the understanding of these processes, the local modulation of the protein levels was studied by using a short KCl stimulation protocol (10 min). This brief depolarisation induces an increase in immunofluorescence for BDNF and TrkB especially in the dendritic compartment (Fig. 3 A, B, C, D). The quantitative analysis demonstrated that a brief depolarisation results in a significant increase in the levels of BDNF and TrkB proteins in the distal parts of the dendritic compartment, and, to a lesser extent, also in the proximal part (Fig. 3 E, F). The increase of the protein levels is completely abolished by the protein synthesis inhibitor cycloheximide (Fig. E, F), demonstrating that it is protein synthesis dependent process.

It is particularly striking to observe that the activity-dependent increase of immunofluorescence for both BDNF and TrkB, due in part to local protein synthesis and in part to protein transport from cell body, is mainly localised in the distal portion of the apical dendrites. The differential enhanced synthesis of BDNF and TrkB in the distal dendrites might be due to selective translation of particular mRNA isoforms that are targeted to the dendrites.

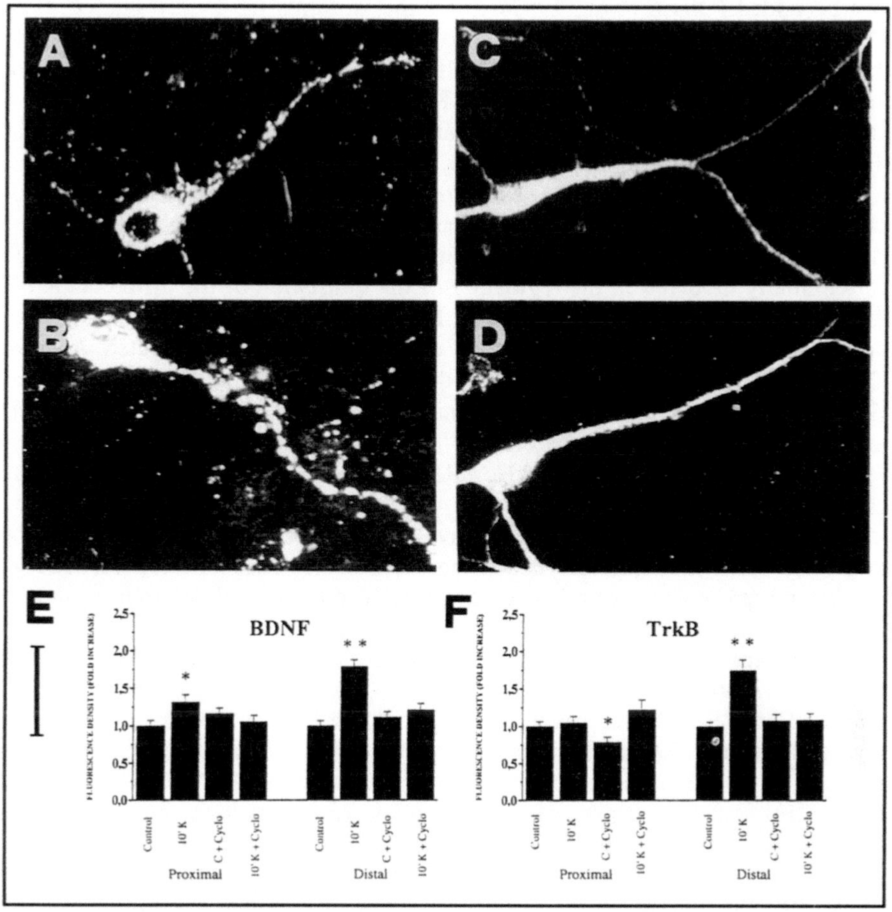

Fig. 3. Immunohistochemistry on cultured hippocampal neurons. A) Anti-BNDF immunostaining of control cultures. B) Anti-BNDF immunostaining after 10 mM KCl depolarization for 10 minutes. C) Staining with anti TrkB antibody in control conditions. D) Staining with anti TrkB antibodies, after 10 minutes depolarization in 10 mM KCl. Fluorescent density of BDNF (A,B) and TrkB (C,D) was determined in proximal and distal regions of the projections of labelled dendrites as described under Material and Methods. Bars represent the mean fold increase of the fluorescence density of 45 dendrites, with respect to the controls (=1.0). Error bars: Standard Error. Quantification of the increase of BDNF (E) and TrkB (F) fluorescence density in both proximal and distal regions (10'K). Incubation of control cells with the protein synthesis inhibitor cycloheximide has no effect (C+10'K). Cycloheximide completely inhibits the increase in fluorescence density induced by the 10 mM KCl stimulus (10'K+Cyclo). Scale bar in (E) is 20μm for A-D. C,E,F reprinted from Tongiorgi et al. (1997) Courtesy of J. Neurosci.

In the mammalian brain, five BDNF mRNA isoforms can be generated by alternative splicing by addition of one of five different exons at the 5' of a common exon containing the coding region (Ohara et al., 1992; Bishop et al., 1994). Moreover to each of these isoforms, two different 3' untranslated regions can be added (Ohara et al., 1992). All these isoforms have been shown to be abundant in the rat hippocampus and to form two different transcripts of either 1.6 Kb or 4.2 Kb in lenght. In addition, it has been shown that all these isoforms are translated into protein but the 1.6 Kb mRNA isoforms, characterised by a short 3' untranslated region, are more enriched in the polysomal fraction than the 4.2 Kb longer transcript (Timmusk et al., 1993; 1994). This finding has suggested a potential translational discrimination for the different isoforms. The different 3' tail is likely to be involved in the differential targeting of the mRNAs towards the dendrites or the cell soma. In addition the different 5' exons might also be differentially involved in the regulation of the translation of the different BDNF mRNAs in response to different stimuli. If this hypothesis is true, the existence of activity-responsive elements in the 5' of these mRNAs is to be expected. Also for TrkB, different isoforms produced by differential splicing at 3' end have been described (Berkemeier et al., 1991; Klein et al., 1991; Middlemas et al., 1991) and more recently, three different 5' alternative spliced exons have been identified at the 5' end of TrkB mRNAs (Benedetti, Tongiorgi, Righi and Cattaneo, unpublished results).

In conclusion, BDNF and TrkB are involved in a series of processes occurring at different time scales, having in common their dependence on electrical activity. The most rapid events occurring in the seconds to minutes time range, comprise the depolarization dependent BDNF release (Goodman et al., 1996) and the rapid synthesis and dendritic accumulation of new BDNF and TrkB proteins (Tongiorgi et al, 1997, and this paper). In the time range of 0.5-3 hours, BDNF and TrkB mRNAs are recruited in an activity-dependent way to the dendrites (Tongiorgi et al., 1997, and this paper). Electrical activity and pharmacologically-induced seizures induce an upregulation of the transcription of BDNF and TrkB genes. However the timing of these events is complex, since the onset of the transcription ranges from 15 minutes from the stimulus, in the case of some BDNF transcripts (Lauterborn et al., 1996) to at least 1-3 hours for other BDNF transcripts and TrkB mRNAs (Zafra et al., 1990; 1992; Elliott et al. 1994; Castrén et al., 1992, Patterson et al., 1992; Castrén et al., 1993; Dragunow et al., 1993; Wetmore et al., 1994; Kokaia et al., 1993, Lauterborn et al., 1996). Moreover, the temporal patterns of synaptic activity and the different temporal phases of synaptic enhancement are important in determining the neurotrophin dependence of plasticity in the hippocampus. Slices treated with anti-TrkB antibodies showed significant deficits in LTP induced by either theta-burst stimulation (TBS) or "pairing" but not by tetanus stimulation. The application of a TrkB-IgG 30 min after LTP induction caused previously potentiated synaptic transmission to return to baseline levels, indicating that TrkB ligands are required to maintain LTP for up to 1 hr after induction. Also the late phase of LTP (2-3 hr) was significantly impaired in slices pretreated with TrkB Ab or a TrkB-IgG indicating that TrkB activation is also involved in sustaining the late, protein synthesis dependent phase of LTP (Kang et al., 1997). In sum, these data reinforce the hypothesis that a tight spatial and temporal regulation of BNDF expression and of its receptor TrkB is instrumental to the establishment of long term synaptic changes. The possibility of a local regulation of the protein synthesis of BDNF and TrkB, possibly through an autocrine dendritic loop, would elegantly explain how spatial and temporal selectivity would occur.

(Experimental work was supported by research grants from Human Frontier Science Project Organisation (HFSPO) (RG93-93), and from MURST (Cofinanziamento'98) to AC).

3 Materials and Methods

3.1 Cell cultures

Primary cell cultures were made from rat hippocampal neurons according to Malgaroli and Tsien (1992), with slight modifications. Hippocampi were dissected from 2-4 day old animals. Isolation and slicing was performed in 200 μM kynurenic acid (Sigma) and 25 μM 2-amino-5-phosphonovalerate (Tocris Neuramin). Tissue slices were digested with trypsin in the presence of DNAse, blocked with trypsin inhibitor on ice, dissociated in medium containing DNAse. Cells were recovered and washed by two successive centrifugations at 500 rpm and plated on glass coverslips coated with 50μg polyornitine and 2% Matrigel (Collaborative Research) in 35 mm Nunc Petri dishes. Cells were cultured for 7 days in 5% CO_2-humidified incubator, in 2 ml minimum essential medium with Earle's salts and Glutamax I (Gibco) additioned with 5-10% fetal bovine serum, 7 mg/ml D-glucose, 3.6 mg/ml HEPES, 0.1 μg/ml biotin, 1.5 μg/ml vitamin B12, 30 μg/ml insulin and 100 μg/ml bovine transferrin. Proliferation of non-neural cells was prevented by the addition of 2.5-5 μM cytosin ß-D-arabinofuranoside from the second day in culture onwards.

3.2 KCl stimulation of cultured hippocampal neurons

Extracellular control solution contained 3.5 mM KCl, 132 mM NaCl, 1 mM $MgCl_2$ 2 mM $CaCl_2$, 20 mM D-glucose, 10 mM Hepes (pH 7.4). For mRNA or protein localization experiments, cells were depolarised for the indicated times, at 37°C, with the K-medium containing 10 mM KCl, 1.8 mM $CaCl_2.2H_2O$, 0.8 mM $MgSO_4.7H_2O$, 101 mM NaCl, 26 mM $NaHCO_3$, 1 mM $NaH_2PO_4.2H_2O$, 0.7% D-glucose, 15 mM Hepes (pH 7.4).

3.3 Riboprobes and oligonucleotides

The rat BDNF cDNA pBCDPst (nucleotides 74-525, Maisonpierre et al., 1991) was kindly provided by Dr. A. Negro (Fidia Res. Lab., Padova). The rat TrkB cDNA clone was kindly provided by Dr. Y Bozzi (Institute of Neurophysiology,

CNR, Pisa) (Bozzi et al., 1995), and contained the first 238 bp of the region coding for the tyrosine-kinase domain (nucleotides 2163-2401, Middlemas et al., 1991). After linearization of the plasmids, the digoxigenin labelled riboprobes were synthesised with a SP6/T7 DIG-RNA labelling kit (Boehringer) according to the manufacturer's instructions.

3.4 In situ hybridisation on cultured hippocampal neurons

For in situ hybridisation with riboprobes, cells were fixed 10 minutes at room temperature in 4% paraformaldeyde in phosphate buffer saline (PBS), washed in PBS and permeabilized in ethanol for 15 minutes at -20°C. After rehydratation with decreasing ethanol concentrations in PBS at r.t., cells were prehybridised at 55°C for 90min in the hybridisation mix containing: 20mM Tris/HCl (pH 7.5), 1mM EDTA, 1x Denhardt's solution, 300mM NaCl, 100 mM dithiothreitol, 0.5mg/ml salmon sperm DNA, 0.5mg/ml polyadenylic acid and 50% formamide. In situ hybridisation was carried overnight at 55°C in the hybridisation mix additioned with 10% dextrane sulfate and the riboprobes (50ng/ml). High stringency washes were carried out in 0.1x sodium saline citrate buffer (SSC)/0.1% Tween20 at 60°C. Cells hybridised with digoxigenin-labelled riboprobes were incubated overnight at 4°C with anti-DIG Fab fragments coupled to alkaline phosphatase (Boehringer), diluted 1:500 in 10% fetal calf serum in PBS additioned with 0.1% Tween20 (PBST). After a thorough wash in PBST, cells were reacted with 4-nitro blue tetrazolium and 5-bromo-4-chloro-3-indolyl-phosphate in 100 mM Tris/HCl (pH 9.5), 50 mM $MgCl_2$, 100 mM NaCl, 1mM Levamisol. Alkaline phosphatase development was carried out for 16 hours at 4°C.

3.5 Antibody staining of cultured hippocampal neurons

Immunohistochemistry on cultured hippocampal neurons was preceded by the same fixation and permeabilisation steps used for in situ hybridisation described above. Fixed and permeabilised cells were preincubated 30 minutes at r.t. in 3% BSA in PBS, incubated 3 hours at r.t. with an antibody recognizing the TrkB full length isoform (Santa Cruz Ab794, made in rabbit, diluted 1:100 in 3% BSA in PBS), or anti BDNF (Promega, made in chicken, diluted 1:100 in 3% BSA in PBS). After washing in PBS, cells were incubated 1 hour at r.t. with biotinylated anti rabbit IgG antibody (Vector) or biotinylated anti chicken IgG antibody (Promega) diluted 1:200 in 3% BSA in PBS, then washed in PBS, incubated in Streptavidin-FITC as described above and mounted in Vectashield (Vector).

3.6 Quantitative imaging analysis and statistics

Non radioactive in situ hybridisation was analysed by viewing stained cultures under bright field illumination with a Zeiss microscope. Stained neurons were acquired with a DAGE-MTI 3CCD colour camera and digitised with the image analysis program Optimas 6.0 (Optimas Corp.). The function "Trace" was used to measure, starting from the base of the dendrites, the maximal distance of dendritic labelling (MDDL). Dendrites were traced, in a conservative manner, up to the point at which the in situ labelling was clearly distinguishable from the background. The background level obtained in sister cell cultures hybridized with the sense probes was used as a reference to distinguish the actual labelling obtained with the antisense probes from the background. The individual preparations were coded and analysed in a blind manner.

Fluorescent immunocytochemistry was analysed by confocal microscopy with a Molecular Dynamics MultiProbe 2001 confocal device mounted on a Nikon microscope. The individual preparations were coded and analysed in a blind manner. Individual neurons whose entire dendritic domains were distinguishable from those of neighbouring neurons were selected. For each neuron, 5 optical sections 0.7 μm thick were acquired and then were integrated in a single projection by the function "look through" to ensure the visualisation of the entire width of the dendrites on the z axis. Quantification of the fluorescence intensity and area measurements were carried out by using the program "Area" of the Molecular Dynamics software packagings. Dendrites, longer than 100 mm, were subdivided into two segments and manually contoured. The proximal segment included the region from the base of the dendrite and the proximal region of the dendrites up to 30 μm. The distal segment comprised the distal region of the dendrites extending from 30 μm to 90 μm from the dendritic base. The sum of the total pixel intensities for each segment was normalised to the area of the segment measured (fluorescence density). The data for the in situ MDDL or the fluorescence density were statistically analysed with unpaired t Student's tests.

4 Literature

Barde Y-A (1990) The nerve growth factor family. Progr Growth Factor Res 2: 237-248.

Berkemeier LR, Winslow JW, Kaplan DR, Nikolics K, Goeddel DV, Rosenthal A (1991) Neurotrophin-5: a novel neurotrophic factor that activates trk and trkB. Neuron 7: 857-866.

Bishop JF, Mueller GP and Mouradian MM (1994) Alternate 5' exons in the rat brain-derived neurotrophic factor gene: differential patternsof expression across brain regions. Brain Reș Mol Brain Res 26: 225-232.

Blöchl A, Thoenen H (1995) Characterization of nerve growth factor (NGF) release from hippocampal neurons: evidence for a constitutive and an unconventional sodium-dependent regulated pathway. Eur J Neurosci 7: 1220-1228.

Blöchl A, Thoenen H (1996) Localization of cellular storage compartments and sites of constitutive and activity-dependent release of nerve growth factor (NGF) in primary cultures of hippocampal neurons. Mol Cell Neurosci 7: 173-190.

Bonhoeffer T (1996) Neurotrophins and activity-dependent development of the neocortex. Curr Op Neurobiol 6: 119-126.

Bozzi Y, Pizzorusso T, Cremisi F, Rossi FM, Barsacchi G, Maffei L (1995) Monocular deprivation decreases the expression of messenger RNA for Brain-Derived Neurotrophic Factor in the rat visual cortex. Neuroscience 69: 1133-1144.

Burgin KE, Waxham MN, Rickling S, Westgate SA, Mobley WC, Kelly PT (1990) In situ hybridization histochemistry of Ca2+/calmodulin-dependent protein kinase in developing rat brain. J Neurosci 10:1788-1798.

Carmignoto G, Pizzorusso T, Tia S, Vicini S (1997) Brain-derived neurotrophic factor and nerve growth factor potentiate excitatory synaptic transmission in the rat visual cortex. J Physiol 498: 153-164.

Castrén E, Zafra F, Thoenen H, Lindholm D (1992) Light regulates expression of brain-derived neurotrophic factor mRNA in rat visual cortex. Proc Natl Acad Sci USA 89: 9444-9448.

Castrén E, Pitkänen M, Sirviö J, Parsadanian A, Lindholm D, Thoenen H, Riekkinen PJ (1993) The induction of LTP increases BDNF and NGF mRNA but decreases NT-3 mRNA in the dentate gyrus. NeuroReport 4:895-898.

Crino PB, Eberwine J (1996) Molecular characterization of the dendritic growth cone: regulated mRNA transport and local protein synthesis. Neuron 17:1173-1187.

Davies AM (1994) The role of neurotrophins in the developing nervous system. J Neurobiol 25: 1334-1348.

Dragunow M, Beilharz E, Mason B, Lawlor P, Abraham W, Gluckman P (1993) Brain-derived neurotrophic factor expression after long-term potentiation. Neurosci Lett 160: 232-236.

Dugich-Djordjevic MM, Peterson C, Isono F, Ohsawa F, Widmer HR, Denton TL, Bennett GL, Hefti F (1995) Immunohistochemical visualization of Brain-derived Neurotrophic Factor in the rat brain. Eur J Neurosci 7:1831-1839.

Elliott RC, Inturrisi CE, Black IB, Dreyfus CF (1994) An improved method detects differential NGF and BDNF gene expression in response to depolarization in cultured hippocampal neurons. Mol Brain Res 26:81-88.

Engert F, Bonhoeffer T (1997) Synapse specificity of long-term potentiation breaks down at short distances. Nature 388:279-284.

Frey U, Morris RG (1997) Synaptic tagging and long-term potentiation. Nature 385: 533-536.

Furuichi T, Samon-Chazottes D, Fujino I, Yamada N, Hasegawa M, Miyaki A, Yoshikawa S, Guenet J-L, Mikoshiba K (1993) Widespread expression of inositol 1,4,5-triphosphate receptor type 1 gene (Insp3r1) in the mouse central nervous system. Receptor Channels 1: 11-24.

Garner CC, Tucker RP, Matus A (1988) Selective localization of messenger RNA for the cytoskeletal protein MAP2 in dendrites. Nature 336: 674-677.

Gazzaley AH, Benson DL, Huntley GW, Morrison JH (1997) Differential subcellular regulation of NMDAR1 protein and mRNA in dendrites of dentate gyrus granule cells after perforant path transection. J Neurosci 17: 2006-2017.

Goodman LJ, Valverde J, Lim F, Geschwind MD, Federhoff HJ, Geller AI, Hefti F (1996) Regulated release and polarized localization of brain-derived neurotrophic factor in hippocampal neurons. Mol Cell Neurosci 7: 222-238.

Herb A, Wisden W, Catania MV, Maréchal D, Dresse A, Seeburg PH (1997) Prominent dendritic localization in forebrain neurons of a novel mRNA and its product, dendrin. Mol Cell Neurosci 8: 367-374.

Johnston DS (1995) The intracellular localization of messenger RNAs. Cell 81: 161-170.

Kang H, Schumann EM (1996) A requirement for local protein synthesis in neurotrophin-induced hippocampal synaptic plasticity. Science 273: 1402-1406.

Kang H, Schumann EM (1995) Long-lasting neurotrophin-induced enhancement of synaptic transmission in the adult hippocampus. Science 267: 1658-1662.

Kang H, Welcher AA, Shelton D, Schuman EM (1997) Neurotrophins and time: different roles for TrkB signaling in hippocampal long-term potentiation. Neuron 19: 653-664

Klein R, Nanduri V, Jing SA, Lamballe F, Tapley P, Bryant S, Cordon-Cardo C, Jones KR, Reichardt LF, Barbacid M (1991) The trkB tyrosine protein kinase is a receptor for brain-derived neurotrophic factor and neurotrophin-3. Cell 66: 395-403.

Kleiman R, Banker G, Steward O (1993) Inhibition of protein synthesis alters the subcellular distribution of mRNA in neurons but does not prevent dendritic transport of RNA. Proc Natl Acad Sci USA 90: 11192-11196.

Knipper M, Leung LS, Zhao D, Rylett RJ (1994a) Short-term modulation of glutamatergic synapses in adult rat hippocampus by NGF. Neuroreport 5: 2433-2436.

Knipper M, da-Penha-Berzaghi M, Blöchl A, Breer H, Thoenen H, Lindholm D (1994b) Positive feedback between acetylcholine and the neurotrophins nerve growth factor and brain-derived neurotrophic factor in the rat hippocampus. Eur J Neurosci 6: 668-671.

Knowles RB, Kosik KS (1997) Neurotrophin-3 signals redistribute RNA in neurons. Proc Natl Acad Sci USA 94: 14804-14808.

Kokaia Z, Bengzon J, Metsis M, Kokaia M, Persson H, Lindvall O (1993) Coexpression of neurotrophins and their receptors in neurons of the central nervous system. Proc Natl Acad Sci USA 90: 6711-6715.

Korte M, Carrol P, Wolf E, Brem G, Thoenen H, Bonhoeffer T (1995) Hippocampal long-term potentiation is impaired in mice lacking brain-derived neurotrophic factor. Proc Natl Acad Sci 92: 8856-8860.

Lauterborn JC, Rivera S, Stinis CT, Hayes VY, Isackson PJ, Gall CM (1996) Differential effects of protein synthesis inhibition on the activity-dependent expression of BDNF transcripts: evidence for immediate-early gene responses from specific promoters. J Neurosci 16: 7428-7436.

Leibrock J, Lottspeich F, Hohn A, Hofer M, Hengerer B, Masiakowski P, Thoenen H, Barde Y-A (1989) Molecular cloning and expression of brain-derived neurotrophic factor. Nature 341: 149-152.

Leßmann V, Gottmann K, Heumann R (1994) BDNF and NT-4/5 enhance glutamatergic synaptic transmission in cultured hippocampal neurones. NeuroReport 6: 21-25.

Levi-Montalcini R. (1987) The nerve growth factor 35 years later. Science 237: 1154-1162.

Levine ES, Dreyfus CF, Black IB, Plummer MR (1995) Brain-derived neurotrophic factor rapidly enhances synaptic transmission in hippocampal neurons via postsynaptic tyrosine kinase receptors. Proc Natl Acad Sci USA 92: 8074-8077.

Lewin GR and Barde YA (1996) Physiology of the neurotrophins. Annu Rev Neurosci 19: 289-317.

Lindholm D, Castrén E, Berzaghi M, Blöchl A, Thoenen H (1994) Activity dependent and hormonal regulation of neurotrophin mRNA levels in the brain implications for neuronal plasticity. J Neurobiol 25: 136-1372.

Lyford GL, Yamagata K, Kaufmann WE, Barnes CA, Sanders LK, Copeland NG, Gilbert DJ, Jenkins NA, Lanahan AA, Worley PF (1995) Arc, a growth factor and activity-regulated gene, encodes a novel cytoskeleton-associated protein that is enriched in neuronal dendrites. Neuron 15: 433-445.

Maisonpierre PC, Le Beau MM, Espinosa RI, Ip NY, Belluscio L, La Monte S, Squinto S, Furth ME, Yancopulos GD (1991) Human and rat BDNF and NT3: gene structures, ditributions and chromosomal localization. Genomics 10: 558-568.

Malgaroli A, Tsien RW (1992) Glutamate-induced long term potentiation of the frequency of miniature synaptic currents in cultured hippocampal neurons. Nature 357: 134-139.

Martone ME, Pollock JA, Jones YZ, Ellisman MH (1996) Ultrastructural localization of dendritic messenger mRNA in adult rat hippocampus. J Neurosci 16: 7428-7436.

Middlemas DS, Lindberg RA, Hunter T (1991) TrkB, a neural receptor protein-tyrosine kinase: evidence for a full-length and twotruncated receptors. Mol Cell Biol 11: 143-153.

Ohara O, Gahara Y, Teraoka H, Kitamura T (1992) A rat brain-derived neurotrophic factor-encoding gene generates multiple transcripts through alternative use of 5' exons and polyadenylation sites. Gene 121: 383-386.

Patterson SL, Grover LM, Schwartzkroin PA, Bothwell M (1992) Neurotrophin expression in rat hippocampal slices: a stimulus paradigm inducing LTP in CA1 evokes increases in BDNF and NT3 mRNAs. Neuron 9: 1081-1088.

Prakash N, Fehr S, Mohr E, Richter D (1997) Dendritic localization of rat vasopressin mRNA: ultrastructural analysis and mapping of targeting elements. Eur J Neurosci 9: 523-532.

Racca C, Gardiol A, Triller A (1997) Dendritic and postsynaptic localizations of glycin receptor a subunits mRNAs. J Neurosci 17: 1691-1700.

Righi M, Tongiorgi E and Cattaneo A (1998) BDNF modulates the dendritic targeting of BDNF and TrkB mRNAs in cultured hippocampal neurons. European Neuroscience Association Abstr. Eur J Neurosci in press.

Schuman E (1997) Synapse specificity and long-term information storage. Neuron 18: 339-342.

Spacek J, Harris KM (1997) Three-dimensional organization of smooth endoplasmic reticulum in hippocampal CA1 dendrites and dendritic spines in the immature and mature rat. J Neurosci 17: 190-203.

Steward O (1994) Dendrites as a compartment for macromolecular synthesis. Proc Natl Acad Sci USA 91: 10766-10768.
Steward O (1997) mRNA localization in neurons: a multipurpose mechanism. Neuron 18: 9-12.
Steward O, Levy WB (1982) Preferential localization of polyribosomes under the base of dendritic spines in granule cells of the dentate gyrus. J Neurosci 2: 284-291.
Thoenen H (1995) Neurotrophins and neuronal plasticity. Science 270: 593-598.
Tiedge H, Brosius J (1996) Translational machinery in dendrites of hippocampal neurons in culture. J Neurosci 16: 7171-7181.
Timmusk T, Palm K, Metsis M, Reintam T, Paalme V, Saarma M, Persson H (1993) Multiple promoters direct tissue-specific expression of the rat BDNF gene. Neuron 10: 475-489.
Timmusk T, Persson H, Metsis M (1994) Analysis of transcriptional initiation and translatability of brain-derived neurotrophic factor mRNAs in the rat brain. Neurosci Lett 177: 27-31.
Tongiorgi E, Righi M, Cattaneo A (1996) Subcellular localization of neurotrophins and neurotrophin receptors: implications for synaptic plasticity. Rev Brasil Biol 56: 175-182.
Tongiorgi E, Righi M, Cattaneo A (1997) Activity-dependent dendritic targeting of BDNF and TrkB mRNAs in hippocampal neurons. J Neurosci 17: 9492-9505
Torre ER, Steward O (1992) Demonstration of local protein synthesis within dendrites using a new cell culture system that permits the isolation of living axons and dendrites from their cell bodies. J Neurosci 12:762-772.
Torre ER, Steward O (1996) Protein synthesis within dendrites: glycosylation of newly synthesized protein in dendrites of hippocampal neurons in culture. J Neurosci 16: 5967-5978.
Wetmore C, Olson L, Bean AJ (1994) Regulation of brain-derived neurotrophic factor (BDNF) expression and release from hippocampal neurons is mediated by non-NMDA type glutamate recptors. J Neurosci 14: 1688-1700.
Wu K, Xu JL, Suen PC, Levine E, Huang YY, Mount HT, Lin SY, Black IB (1996) Functional trkB neurotrophin receptors are intrinsic components of the adult brain postsynaptic density. Brain Res Mol Brain Res 43: 286-290.
Zafra F, Castren E, Thoenen H, Lindholm D (1991) Interplay between glutamate and gamma-aminobutyric acid transmitter systems in the physiological regulation of brain-derived neurotrophic factor and nerve growth factor synthesis in hippocampal neurons. Proc Natl Acad Sci USA 88: 10037-10041
Zafra F, Hengerer B, Leibrock J, Thoenen H, Lindholm D (1990) Activity-dependent regulation of BDNF and NGF mRNAs in the rat hippocampus is mediated by non-NMDA glutamate receptors. EMBO J 9: 3345-3550.
Zafra F, Lindholm D, Castren E, Hatikka J, Thoenen H (1992) Regulation of brain-derived neurotrophic factor and nerve growth factor mRNA in primary cultures of hippocampal neurons and astrocytes. J Neurosci 12: 4793-4799.

Map projection rewiring in the adult cerebellum after lesions

P. Strata, M. Zagrebelsky, M. Bravin and F. Rossi

Department of Neuroscience, University of Turin, Corso Raffaello 30, I-10125, Turin, Italy

Abstract. Functional recovery following a lesion in the brain depends on the ability of the injured neurones to regrow their axons up to the target and to restore the original projection map. In this review we show that these two important processes occur in the rat olivocerebellar pathway both during the developmental stage and in the mature brain. Axotomised olivocerebellar neurones regenerated into a graft of freshly dissociated Schwann cells and some of them reached previously deafferented Purkinje cells. To study the specificity of the newly formed connections we have adopted three experimental protocols aimed at seeing the specificity of reinnervation into sagittally oriented Purkinje cell bands identified by zebrin II antibodies. First, during postnatal development, following unilateral pedunculotomy, fibres from the uninjured pathway cross the cerebellar midline to reinnervate symmetrical Purkinje cell bands. Second, by grafting embryonic Purkinje cells in the mature cerebellum, olivocerebellar fibres reached the specific neurochemically defined Purkinje cell compartments into the graft. Third, in the mature cerebellum, following a subtotal inferior olive lesion, collateral sprouting of the surviving climbing fibres was confined into the proper cortical compartment. These results show that olivocerebellar fibres have a strong intrinsic capability to regrow their axons and that this growth is not randomly oriented, but it is regulated by selective interactions between distinct subsets of inferior olive axons and target neurones.

Keywords. olivocerebellar system, climbing fibre, axon regeneration, reinnervation, topography, neural transplantation

1 Introduction

Neurones in the central nervous system (CNS) may be classified into two broad categories. One is formed by highly branched axons, projecting to a large number of target neurones without a precise topography. The other is characterised by axonal projections arranged according to topographically ordered maps within their target regions. Highly sophisticated functions of the brain, like those of sensory and motor systems, are strictly dependent on these topographical maps.

Thus, an important requirement for achieving a complete functional recovery after brain lesions, is the possibility of reestablishing the original projectional maps.

In the adult mammalian CNS, functional recovery following a lesion is hampered by the lack of nerve cell replication and by minimal ability of axons to elongate. The latter factor is one of the major and most investigated issues in the study of postlesional plasticity. The extent of axonal growth is influenced by intrinsic properties of individual neurones (Fawcett, 1992) as well as by the microenvironment, through which axons should grow. In fact, both the application of antibodies against myelin-associated inhibitors of axonal growth and the grafting of peripheral nerves allow some central axotomised neurones to elongate their axons to some extent (Caroni & Schwab, 1988b; Aguayo, Bray, Rasminsky, Zwimpfer, Carter & Vidal-Sanz, 1990). In these conditions, axons are able to form normal synapses, at ultrastuctural and functional level, with their original target neurones (Keirstead, Rasminsky, Fukuda, Carter, Aguayo & Vidal-Sanz, 1989). In some cases, in the absence of any choice, synapses may also be formed with other neurone types (Zwimpfer, Aguayo, & Bray, 1992). However, when given the possibility to choose they always prefer their original target (Harvey, Gan & Pauken, 1987; Bähr & Eschweiler, 1991; Hankin & Lund, 1991).

A further much less investigated problem is to know whether in mammals, once the new axons have been successful in reaching their target region, they are also able to recognise specific cells in order to reestablish the original map, recapitulating what happens during development. This implies the expression or the reexpression of specific recognition cues by target neurones. In *in vitro* conditions, Wizemann, Thies, Klostermann, Bonhoeffer & Bähr (1993) have shown that when an embryonic or an adult retina is placed in front of membrane preparations of alternate anterior and posterior regions of the adult rat superior colliculus, retinal axons from temporal and nasal retina specifically grow onto the proper membrane stripe. This high specificity is present only when membranes are prepared from the adult superior colliculus at least 12 days following deafferentation. This result suggests that the molecular cues guiding the axon growth toward the appropriate target neurones are reexpressed in the adult colliculus or that repulsive molecules, masking the guiding cues, are removed as a consequence of the lesion. Myelin inhibitory proteins for axonal growth disappear following colliculus deafferentation and loss of these molecules might prevent a correct target selection by the growing axons. However, the latter possibility has been disproved since application of neutralising antibodies (IN-1) to the myelin associated inibitors does not disturb target-specific axon guidance *in vitro* (Bähr & Schwab, 1996).

Formation of projectional maps during development is not confined to positional chemical guidance cues but, in *in vivo* conditions, it most likely involves complex mechanisms. Cell-cell interactions, glial factors, trophic substances and activity-dependent processes, in fact, play an important role in the fine tuning between the presynaptic and the postsynaptic elements. Therefore, also during regeneration in *in vivo* conditions, the reestablishment of the original maps likely depends on multifactorial mechanisms.

In a series of recent experiments performed in our laboratory we faced this problem *in vivo* to establish whether axonal growth following injury is associated with the restoration of organised projectional maps. To address this issue, we selected the olivocerebellar system in the rat, as most suitable for the purpose. In fact, this pathway is endowed with a high degree of plasticity in the terminal arborisation (Rossi & Strata, 1995) and it maintains in the adult the expression of a growth-associated molecule, B-50/GAP-43 (Kruger, Benotti, Rivolta & Samanin, 1993), a protein known to be important for neuronal plasticity, which is also present in uninjured olivocerebellar axons and climbing fibres (Buffo, Fronte, Ostereicher & Rossi, 1998). Moreover, this system has a precise point-to-point organisation (Azizi & Woodward, 1987). During the developmental period, chemospecific interactions occur between olivocerebellar axons and Purkinje cells, their main target in the cerebellar cortex, leading to the establishment of the mature projectional map (Wassef, Angaut, Arsenio-Nunes, Bourrat & Sotelo, 1992). In the adult rat, clusters of inferior olive neurones project to well defined longitudinal Purkinje cell compartments in the contralateral cerebellar cortex, although nearby clusters do not project to adjacent compartments. These compartments may be recognised by a number of chemical cues (for a review, see Herrup & Kuemerle, 1997). Among them, Zebrin II clearly defines seven positive Purkinje cells bands separated by seven negative ones (Hawkes & Herrup, 1995).

2 Olivocerebellar fibres regenerate across a graft of dissociated cells from newborn sciatic nerve and reach their target

In mammals, regeneration of axotomised central neurones is limited by the lack of intrinsic determinants (Fawcett, 1992) and/or by inhibitory influences of the environment (Schwab, Kapfhammer & Bandtlow, 1993). However, some neurones are able to elongate their axons after axotomy, when provided a proper environment, whereas others are not (Rossi, Jankowski & Sotelo, 1995). In a now classical series of investigations (Aguayo, Bray, Rasminsky, Zwimpfer, Carter & Vidal-Sanz, 1990), it has been shown that some central axons can regenerate when connected to a segment of grafted sciatic nerve. In addition, the regenerated axons may reinnervate the deafferented target cells (Keirstead, Rasminsky, Fukuda, Carter Aguayo & Vidal-Sanz, 1989). Schwann cells have thus been used to allow central axons to regenerate, overcoming the effect of a glial scar and providing the necessary substrate for axonal elongation (see Guénard, Xu & Bunge, 1993 for references).

In our experiments (Bravin, Savio, Strata & Rossi, 1997), we grafted freshly dissociated sciatic nerve cells from newborn rats (P0-P2) into the cerebellar lesion site to induce severed olivocerebellar axons to elongate and to reach previously deafferented Purkinje cells. As shown in Fig. 1a, a thin blade 2 mm wide was inserted into the dorsal vermis near the midline, with a downward inclination of about 45 degrees, to interrupt olivocerebellar axons. The graft was implanted within the lesion cavity. To label olivocerebellar axons we injected into the inferior olive an anterograde tracer, biotinylated dextran amine (BDA). After a variable period of time from 4 days to 13 weeks, the rats were sacrificed and the cerebella processed for histological examination. Purkinje cells were immunostained with an anti-calbindin antibody, whereas to visualise the Schwann cells we used an antibody specific for the P75 low affinity NGF receptor.

Grafted Schwann cells completely filled the lesion cavity where they often appeared as longitudinally oriented bundles along the lesion track. In addition, they migrated radially and some of them were found isolated throughout the cerebellar parenchyma up to the cerebellar cortex. As shown in Figs. 1b and 1c, olivocerebellar fibres were present throughout the extension of the Schwann cell graft and their course depended on the Schwann cell disposition. It is thus clear that axotomised olivocerebellar fibres were able to overcome the effect of the glial scar at the lesion site and to regenerate across the Schwann cell graft. However, their further growth towards the deafferented cerebellar cortex depended on the local structure the regenerating fibres had in front. When the new fibres were in contact with the white matter, no further growth was ever observed (Fig. 1c). In contrast, when they met the granular layer of the cerebellar cortex, the pattern of growth changed abruptly in a branching mode with locally oriented terminals (Fig. 1c). This picture suggests that the deafferented granule cell layer might be the target of the new fibres, thus preventing their further growth towards the Purkinje cells. It should be emphasised that in normal development, the olivocerebellar fibres meet the Purkinje cells before the granular layer is formed. In addition, collaterals of olivocerebellar fibres are described to end on granule and Golgi cells (Scheibel & Scheibel, 1954; Palay & Chan-Palay, 1974). Therefore, the granule cell layer likely provides the new fibres with natural target cells, thus preventing them to grow further. In addition, it is possible that myelin-associated inhibitors are present in the deafferented granule cell layer, despite the degeneration of the mossy fibres. In fact, Purkinje cells survive to axotomy. Despite the tendency of olivocerebellar fibres to branch and terminate in the granule layer, a few of them were able to grow further and to reach the Purkinje cells (Figs. 1c). Their pattern of innervation was different from that occurring in the normal cerebellum. In fact, the new fibre, although emerging from the granule cell layer, reached the Purkinje cell dendrites not by climbing from the cell body to the proximal dendrites, but by approaching them directly and then elongating further towards dendritic trees nearby located. A further pattern of innervation of the Purkinje cell was seen when the Schwann cell graft was in contact with the surface of the cerebellar cortex. The olivocerebellar fibre, in this case, proceeded from the surface down to the Purkinje cell dendrites where it ended with a terminal arbor similar to a normal climbing fibre.

In conclusion, this series of experiments shows that axotomised olivocerebellar fibres are able to grow across a bridge of acutely dissociated Schwann cells grafted into the lesion site, to expand further their growth predominantly within the granular layer, and sometimes to reach Purkinje cells.

Once established that the olivocerebellar fibres have the intrinsic property for regeneration and that, provided with a permissive environment, they are able to

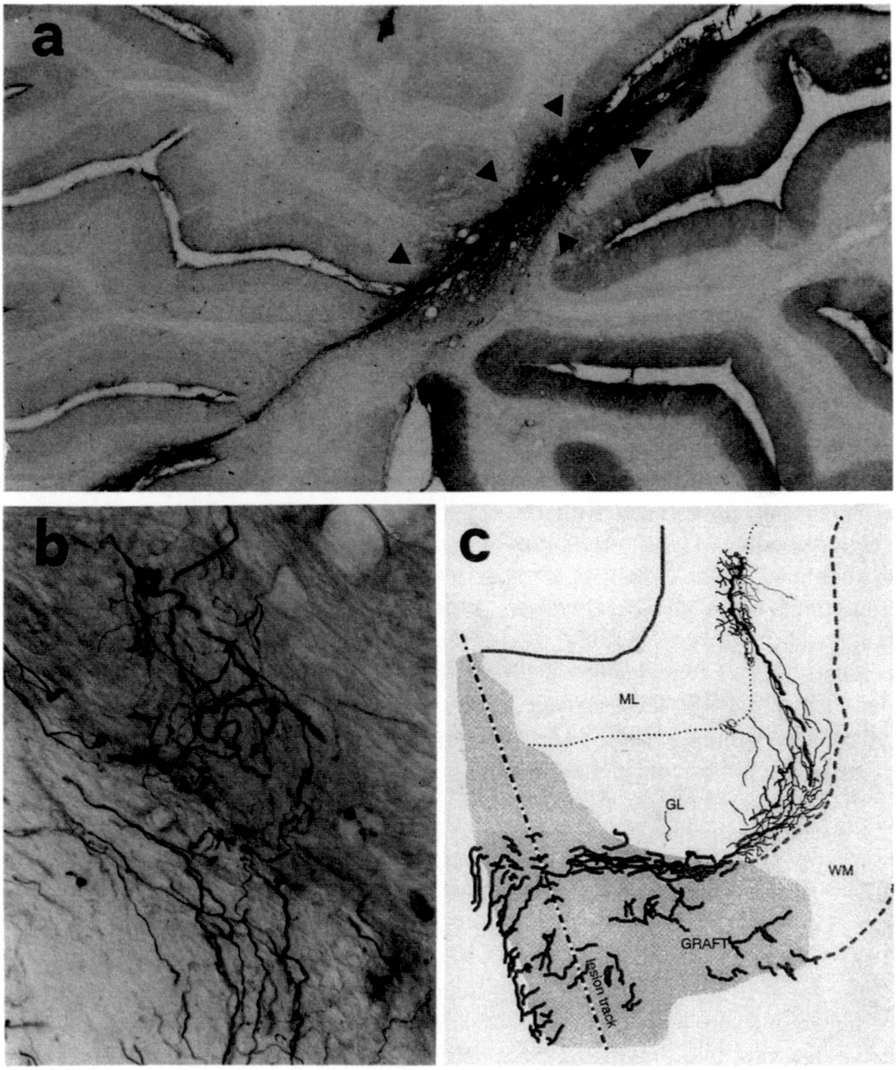

Fig. 1. (a) General features of the cerebellar lesion and Schwann cell transplant in a cerebellar section stained for anti-P75 low affinity NGF receptor. The lesion cavity is partly filled by a Schwann cell transplant (arrowheads) accommodated along the injury track. (b) The micrograph shows several BDA labelled oliovocerebellar axons running across the host-graft interface (dotted line, g = graft). Note the winding course taken by most of these axons within the graft. (c) Camera lucida drawing obtained by superimposing several adjacent sections. The lesioned axons cross the lesion site and run in a bundle across the graft. As they enter the granular layer they break into numerous thin branches which are always excluded from the white matter. Only one thick fibre enters the molecular layer where it is innervating few Purkinje cells.

reinnervate the target, the next question was whether the regenerating fibres are able to reestablish the original projection map. To this aim, we have adopted experimental protocols which confronted the olivocerebellar axons with the Purkinje cells in three different conditions. First, the two neurone types were in the developmental stage. Second, adult olivocerebellar axons interacted with embryonic Purkinje cells. Third, both neurones were in the adult stage.

3 Olivocerebellar projection map formed by compensatory transcommissural reinnervation during development

Angaut, Alvarado-Mallart & Sotelo (1985) have shown that following a complete unilateral section of the inferior cerebellar peduncle, performed at birth, axons from the uninjured contralateral olivocerebellar pathway cross the cerebellar midline and innervate the deafferented hemicerebellum. This newly formed projection is organised in parasagittally oriented strips of climbing fibres symmetrically distributed with respect to their counterparts in the contralateral hemicerebellum. These results provided a first indication of the existence, in the deafferented hemicerebellum, of specific positional cues driving the growth of transcommissural fibres. However, it was still unclear whether the original relationship between distinct olivocerebellar axon subsets and Purkinje cell compartments is reestablished and how the targeting accuracy is achieved during the early phases of development of the transcommissural projection. Thus, we have addressed these issues by several experimental approaches.

First, we have studied the behaviour of the calcitonin gene-related peptide (CGRP)-immunoreactive climbing fibres during the early phases of reinnervation. During development a subset of inferior olivary neurones is characterised by the transient expression of this marker and their axons and terminal arborisations are distributed in the cerebellar cortex according to a precise pattern of parasagittally oriented bands (Kubota, Inagaki, Shimada, Kito, Zaidi, Girgis, MacIntyre & Tohyama, 1987; Morara, Provini & Rosina, 1989; Chédotal & Sotelo, 1992). Thus, we have performed a unilateral complete section of the inferior cerebellar peduncle in P1 rat pups and we have assessed whether the CGRP-immunoreactive axons growing into the deafferented hemicerebellum were able to reestablish the correct distribution pattern in the cerebellar cortex. Already 18-24 hours after the lesion fine terminal branches and labelled axons gradually emerged from the intact medial vermal band and spread across the cerebellar midline in the territories left vacant by the degenerated fibres. These axons and terminal branches gradually covered the whole area formerly occupied by the contralateral band, being strictly confined within its boundaries (Fig. 2a). During the next few days, two additional CGRP-immunoreactive climbing fibre bands appeared more laterally in the vermal cortex, which were clearly separated from one another and symmetrically disposed with respect to those in the uninjured hemicerebellum (Fig. 2a). These results clearly show that a pattern of CGRP-immunoreactive climbing fibre bands is

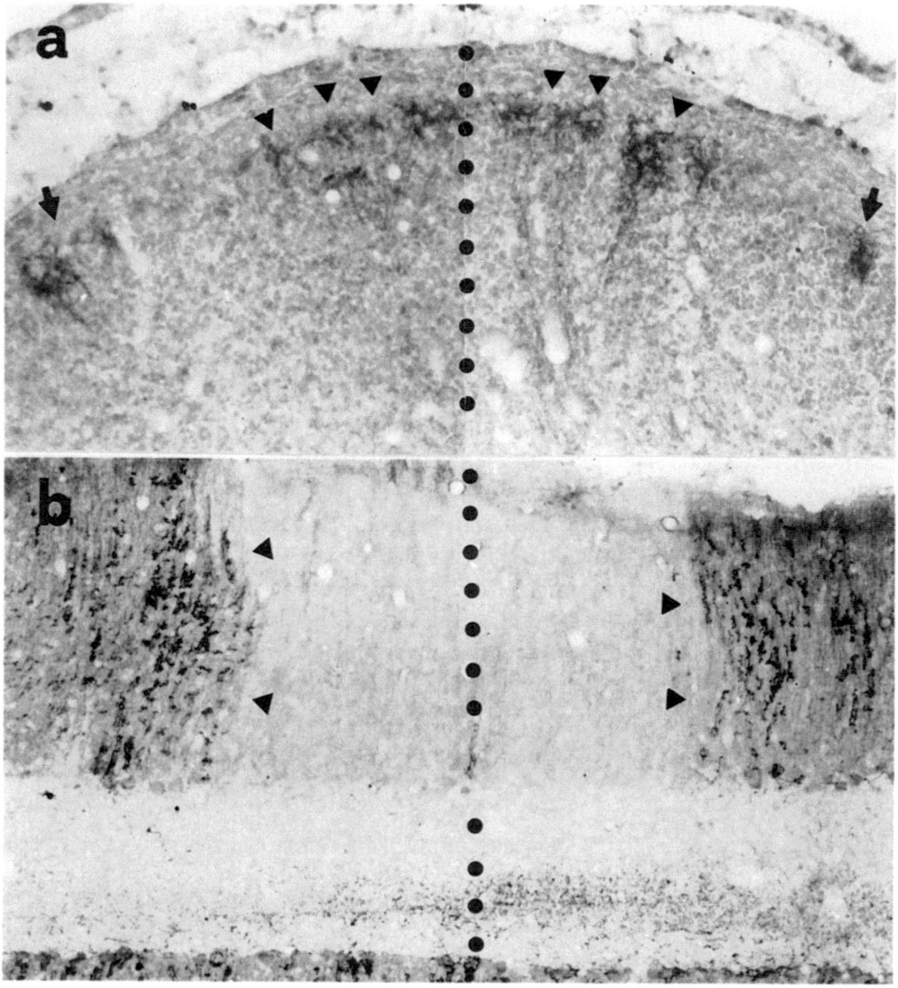

Fig. 2. (a) Reestablishment of the CGRP-immunolabelled climbing fibre pattern in the deafferented hemicerebellum. The micrograph taken six days after the lesion, shows two climbing fibre bands (arrowheads) abutting the cerebellar midline (dotted line) and two other compartments (arrows) situated more laterally. Note the symmetrical disposition of the labelled axons in the two hemicerebella. (b) The low power micrograph shows the medial region of the posterior vermal cortex. Note the highly symmetrical disposition of BDA-labelled climbing fibre strips with respect to the cerebellar midline (dotted line). In addition, two zebrin II-defined Purkinje cell compartments are partially covered by labelled climbing fibres whose terminal arbours abruptly stop at the band border (arrowheads) indicating that the strict alignement between climbing fibre strips and zebrin-defined Purkinje cell compartments is reestablished.

reestablished in a very precise manner within few days after the lesion (Zagrebelsky, Strata, Hawkes & Rossi, 1997).

In normal rat cerebella anti-zebrin II antibodies reveal alternating positive and negative Purkinje cell compartments forming a precise pattern of parasagittally oriented bands which are precisely aligned to climbing fibre strips (Hawkes, Colonier & Leclerc, 1985; Gravel, Eisenman, Sasseville & Hawkes, 1987; Brochu, Maler & Hawkes, 1990). Thus, in adult rats pedunculotomised at birth (P1), we assessed whether the newly formed climbing fibre bands align to the zebrin II-defined Purkinje cell compartments. The relationship between the terminal fields of olivocerebellar axons and zebrin II-identified Purkinje cell bands was examined in frontal cerebellar sections, in which BDA tracing was used to visualise climbing fibres and zebrin II immunocytochemistry to stain Purkinje cell compartments. Since BDA injections did not cover the whole inferior olive, only a subpopulation of olivocerebellar axons was traced in the cerebellum. In the deafferented hemicerebellum the labelled climbing fibres terminate as several parasagittal strips, symmetrically organised with respect to those in the intact side. Furthermore, climbing fibre strips were aligned to zebrin II +/– Purkinje cell compartments and they sharply ended at their borders (Fig. 2b). In addition, also the distribution of the olivocerebellar axon collaterals in the deep cerebellar nuclei was symmetrical to that observed in the intact side. Thus, these results show that the newly formed transcommissural projection is able to restore position-specific connections with target neurones in the cerebellar cortex and deep nuclei.

During normal development, the olivocerebellar axons are segregated into distinct subsets before they reach their target (Chédotal & Sotelo, 1992). Then, each axon subset meets the corresponding population of Purkinje cells to establish a precise projectional map (Sotelo, Bourrat & Triller, 1984; Chédotal & Sotelo, 1992). Our results show that the transcommissural reinnervation process is able to recapitulate these phenomena, although in this condition axon outgrowth occurs along an aberrant pathway across the cerebellar midline and at a later developmental stage. Thus, the results obtained indicate that transcommissural olivocerebellar reinnervation is not random, but it is regulated by selective interactions between distinct subsets of olivocerebellar axons and target neurones aimed at reestablishing the correct projection map.

4 Olivocerebellar fibres innervate specific subsets of embryonic Purkinje cells grafted intraparenchymally into the adult cerebellum

This series of experiments was planned to study whether regenerating olivocerebellar fibres in the adult stage were able to recognise specific subsets of embryonic Purkinje cells. To investigate this, we grafted embryonic cerebellar anlage, taken at E14, intraparanchymally into an adult cerebellum. Olivocerebellar fibres were labelled by means of BDA and Purkinje cells were visualised either with anti-zebrin II and/or anti-calbindin antibodies.

As shown by Rouse & Sotelo (1990), Purkinje cells in the graft developed a mature phenotype and they tended to form clusters of zebrin II+ and zebrin II- neurones. In addition it has been shown that olivocerebellar fibres were able to grow inside the graft to form functional synapses with Purkinje cells (Armengol, Sotelo, Angaut & Alvarado-Mallart, 1989).

In our experiments we aimed at seeing whether there was any selectivity in the innervation of the two populations of Purkinje cells defined by the zebrin II immunoreactivity. Since BDA injections did not fill all regions of the inferior olive, only a subpopulation of olivocerebellar axons was traced in the cerebellum. Out of 29 injected rats, 3 animals were selected because the labelled climbing fibres were distributed almost entirely onto the zebrin positive compartments of the intact cortex. In these rats we analysed the distribution of the BDA labelled fibres into positive and negative zebrin islands inside the graft. As shown in Fig. 3, labelled fibres were highly represented inside the zebrin positive islands.

These results strongly suggest that adult regenerating olivocerebellar fibres are endowed with the capability of recognising specific populations of embryonic Purkinje cells.

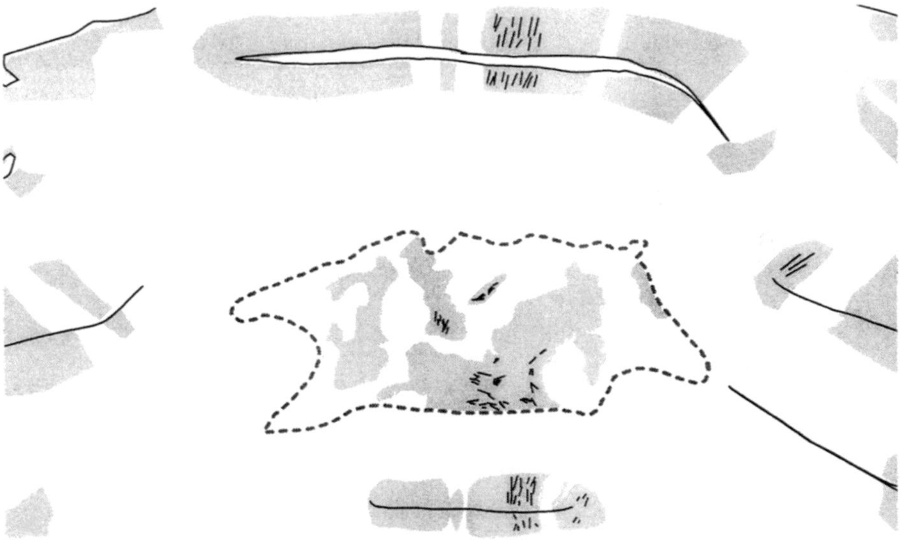

Fig. 3. Camera lucida reconstruction of a frontal cerebellar section containing the embryonic cerebellar graft (delineated by the dotted line). Shaded areas represents zebrin II-positive Purkinje cell bands within the recipient cerebellar cortex and zebrin II-positive Purkinje cell regions within the graft. Note that labelled climbing fibres are strictly confined into zebrin II-positive compartments both within the cerebellar cortex and the graft.

5 Topographically organised climbing fibre sprouting in the adult cerebellum

In this series of experiments we tested whether a selectivity in target recognition was still present when both climbing fibres and Purkinje cells were in the adult stage. Following subtotal lesions of the inferior olive in the adult rat, surviving olivocerebellar fibres, scattered throughout the cerebellar cortex, give rise to collateral sprouts that were going to innervate the deafferented Purkinje cells (Benedetti, Montarolo, Strata & Tosi, 1983; Rossi, Wiklund, van der Want & Strata, 1991a,b). We have used this experimental model to show whether this type of collateral sprouting was randomly oriented or instead occurred within the boundaries of the zebrin II compartments.

In the intact rat cerebellum, each inferior olive neurone gives rise, on average, to 7 axonal branches which are going to innervate with their climbing fibres, the same number of Purkinje cells. The branching usually occurs deep in the white matter and the different branches are distributed to cerebellar regions which might be in different lobuli. Only occasionally, a branching is detectable in the white matter just below or in the granular layer, however, it never occurs inside the molecular layer. Only after a subtotal lesion of the inferior olive, collaterals of climbing fibre arbours are detectable in the molecular layer, where they may be seen extending both in the longitudinal extension of the folium or across it (Rossi, Wiklund, van der Want & Strata, 1991a,b).

New collateral sprouts were identified as climbing fibre branches which elongate into the molecular layer and formed terminal arbours on nearby Purkinje cell dendrites. In 8 lesioned rats we have observed 116 of such sprouts. We found that 112 (97%) of them were confined entirely within a zebrin positive compartment (Fig. 4a), whereas 4 (3%) of them crossed the boundary between two zebrin bands (Fig. 4b). However, in all 4 instances the axonal extension was limited to one Purkinje cell dendritic arbour. More significantly, 24 sprouted fibres (21%) stopped at the edge of the compartment or extended along its border (Fig. 4c). We calculated the mean length of sprouted fibres along single frontal sections, which was 54.8 ± 15.2 µm. If the sprouted fibres were growing in a random manner, the distance between the end of a sprouted fibre and the border of a compartment should be distributed with equal probability within the 50 µm around the border. Instead, as shown in Fig. 4d, there was a highly significant number of fibres ending near the band limit and a remarkable number of fibres ending just at this border. Therefore, we conclude that Purkinje cells, previously deafferented of climbing fibres, express the positional cues which are necessary for target recognition and for restoring the projectional map (Zagrebelsky, Rossi, Hawkes & Strata, 1996).

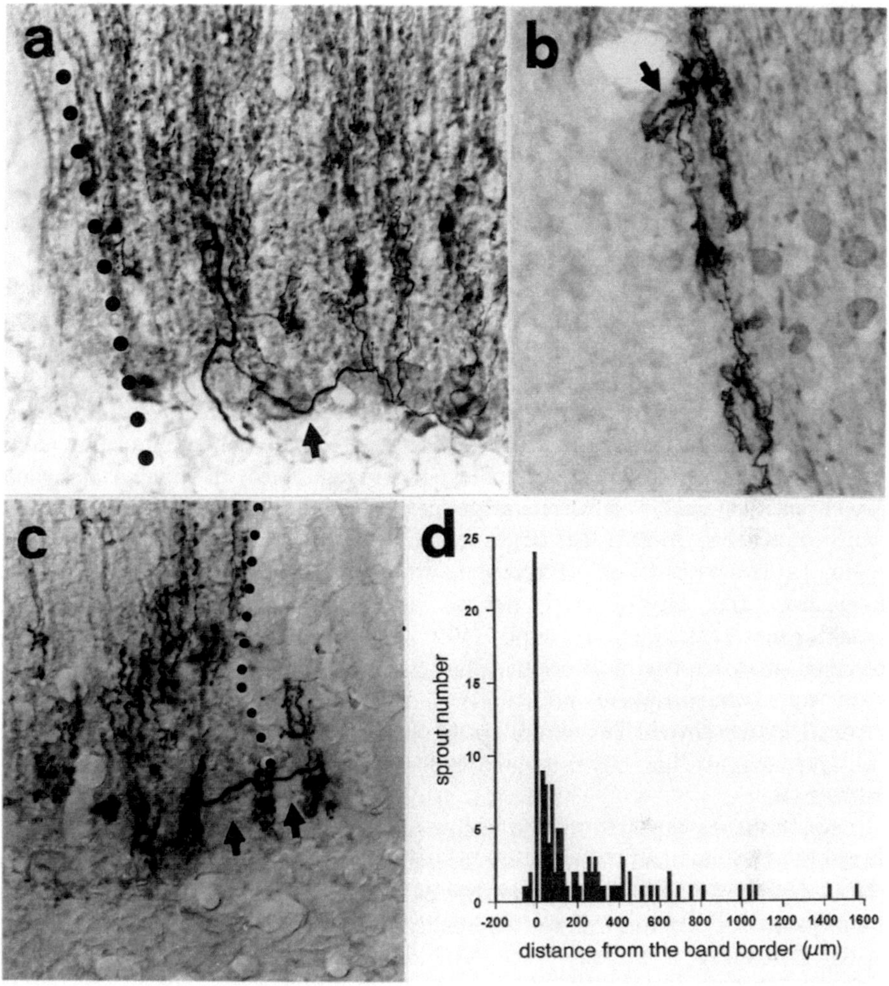

Fig. 4. (a) A climbing fibre collateral (arrow) extends mediolaterally within the molecular layer and innervates neighbouring Purkinje cells. This arbour entirely confined within the zebrin II-positive Purkinje cell band, does not abut the band border (dotted line). (b) A sprouted climbing fibre (arrow) innervates neighbouring Purkinje cells delineating the edge of the zebrin II-positive band without entering the adjacent compartment. (c) A sprouted axon contacts several zebrin II-positive Purkinje cells. In addition, however, it crosses the band border (dotted line) and innervates one zebrin II-negative Purkinje cell (arrows). (d) The distance from the sprouted arbour to the nearest Purkinje cell compartment border is plotted. Positive values belong to arbours confined within a zebrin II-defined band, zero values indicate arbour that directly abut the compartment boundary and negative values indicate the few cases where collateral cross the band border. The high number of sprouts at zero indicates that neurite expansion is preferentially arrested at the Purkinje cell module borders.

6 Discussion

6.1 Regeneration of the olivocerebellar pathway and target reinnervation

The adult olivocerebellar fibres, in addition to the ability to largely modify their terminal arbour extension (Rossi, Wiklund, van der Want & Strata, 1991a,b; Rossi & Strata, 1995; Bravin, Morando, Vercelli, Rossi & Strata, 1996), are able to regenerate their sectioned axons when provided with a permissive environment of acutely dissociated cells from newborn peripheral nerves. Once the regenerated fibres have crossed the graft, their growth could continue only within the cerebellar cortex. Most of them met the granular layer, where their growth changed from an elongating to a branching mode. Evidence coming mainly from *in vitro* studies suggests that during normal development, axonal elongation occurs under the drive of substrate-bound molecules (short-range cues) distributed along the growth pathway and by diffusible target-derived trophic factors (long-range cues) which direct them towards the target region. When fibres have reached the target region, a down-regulation of receptors for local cues prevents further axonal elongation and target-derived factors induce signals for branching and synaptogenesis (Hankin & Lund, 1991). Our experiments suggest similar mechanisms in *in vivo* postlesional plasticity. Axotomised olivocerebellar fibres grow across the peripheral graft under the guidance of local cues towards the cerebellar parenchyma. The abrupt change in their growing mode into the granule cell layer suggests that target-derived factors are present here to modify growing instructions.

It was rather unexpected to find such a signal in the granular layer, since this structure is not the main target of olivocerebellar axons, although their collaterals end in contact with granule cells that were deprived of their normal afferents as a consequence of the surgical deafferentation performed in the underlying white matter. Therefore, it is likely that a large amount of target-derived molecules is released, imparting to the growing olivocerebellar fibres the necessary instructions to change the pattern of growth. It is interesting that when the activity of the adult cerebellar cortex is depressed by a chronic infusion of tetrodotoxin, climbing fibre terminal arbours become atrophic, but they give rise to new collaterals directed towards the granule cell layer. In this case, the new fibres elongate with little branching and they never reach the deeper part of the layer (Bravin, Morando, Vercelli, Rossi & Strata, 1996). Thus, we suggest that inactivity of granule cells and/or of their afferents allow the granule cells to release a tropic factor for the climbing fibres.

It is significant that, despite the granule cell layer forming a barrier to axonal elongation, some olivocerebellar axons could elongate up to the climbing fibre-deprived Purkinje cells, where they could form a typical synapse. This means that the olivocerebellar axons are endowed with the potentiality to achieve not only successful regeneration, but also target reinnervation.

6.2 Specificity of reinnervation

During development, growing axons point towards their target region under the influence of long-range cues. Once they have entered this region, in point-to-point systems, the axons are able to select among different neuronal subpopulations in order to build a topographically ordered map. It is likely that adequate interactions between specific presynaptic and postsynaptic cues are necessary to promote a correct target recognition (Sperry, 1963). The mechanism by which afferent fibres grow to their target have been widely investigated in both submammalian and mammalian species (Sperry, 1963; Jacobson, 1991). In many systems, growing fibres are capable of sensing the direction to their target from a distance, even after experimental misrouting (Stuermer, 1988). From these studies it clearly emerges that during development, both local cues along the path of axonal elongation and target-derived molecules are involved in building the framework of the projectional map. In addition, several lines of evidence suggest that neuronal activity plays an important role in the fine tuning of neuronal connections.

A basic issue in postlesional plasticity of the brain is whether similar mechanisms of target recognition are maintained in the mature brain. While in lower vertebrates the possibility for regeneration and map building is clearly maintained to some extent (Rankin & Cook, 1986; Stuermer, 1988; Jacobson, 1991), in the mammalian brain, regeneration of axotomised axons is hampered by intrinsic and extrinsic factors (Caroni & Schwab, 1988a; Aguayo, Bray, Rasminsky, Zwimpfer, Carter & Vidal-sanz, 1990; Fawcett, 1992). Thus, the study of map restoration following brain lesions in mammals has been poorly investigated, particularly under *in vivo* conditions.

An ordered topography is restored in the adult mammalian peripheral nervous system. In adult rats, for instance, motor axons can reinnervate muscles with the proper topography to some degree (Laskowski & Sanes, 1988) and in the superior cervical ganglion, presynaptic regenerated fibers give rise to a pattern of innervation which is similar to that of the normal ganglion (Njå & Purves, 1977). In the central nervous system, despite the constraint to axonal growth, regenerated or sprouted fibers are able to reinnervate neurons homo- and heterotypically. In *in vivo* conditions, for instance, retinal axons regenerated along a sciatic graft can innervate cerebellar neurons (Zwimpfer, Aguayo & Bray, 1992). In addition, new fibers can innervate their proper target, but on a different dendritic region. However, the new connections are displaced upon the arrival of the original input. Following a lesion of the inferior olive, parallel fibers give rise to collateral sprouting that innervates the proximal dendrites of the Purkinje cell, a region normally occupied by the climbing fibers. However, collaterals of climbing fibers surviving after a subtotal inferior olive lesion, regain their proper territory, thus displacing the abnormal connections (Rossi, Wiklund, van der Want & Strata, 1991a,b). Thus, regenerating axons are clearly able to recognize and contact their proper target cells when these are available, while they can form abnormal connections if they are not given any other choice.

An important issue in postlesional plasticity of the mature brain is whether projectional maps can be restored. To our knowledge, one study clearly demonstrates the presence of positional cues driving regenerating fibers into the proper specific subpopulation of target neurons (see above, Wizenmann, Thies, Klostermann, Bonhoeffer & Bähr, 1993). This study on the retino-tectal projections has been performed under simplified *in vitro* conditions, where growing axons were placed on a proper substrate in front of membrane preparations. *In vivo*, a rough topographical organization in the hamster tectum is achieved by retino-tectal fibers regenerating along a sciatic nerve graft (Thanos, Naskar & Heiduschka, 1997). Our *in vivo* experiments show that a precise topographical projection can be restored in the postnatal period (Zagrebelsky, Strata, Hawkes & Strata, 1997), when adult olivocerebellar fibers are opposed to an embryonic target (Rossi, Fronte, Hawkes & Strata, 1997) and in the mature cerebellum (Zagrebelsky, Rossi, Hawkes & Strata, 1996). These experiments suggests that those specific interactions between presynaptic and postsynaptic partners which guide the building of the projectional map during development (see also Chédotal, Bloch-Gallego & Sotelo, 1996) are also operative in the mature brain. The molecular mechanisms that drive this map reconstruction are unknown. However, the fact that tetrodotoxin treatment of the mature cerebellum reversibly affects the climbing fiber to Purkinje cell connections underlines the importance of electrical activity in this reconstructive process (Bravin, Morando, Vercelli, Rossi & Strata, 1996).

7 Bibliography

Aguayo A.J., Bray G.M., Rasminsky M., Zwimpfer T., Carter D. & Vidal-Sanz M. (1990). Synaptic connections made by axons regenerating in the central nervous system of adult rats. *Journal of Experimental Biology* **153**,199-224.

Angaut P., Alvarado-Mallart R.M. & Sotelo C. (1985). Compensatory climbing fibre innervation after unilateral pedunculotomy in the new born rat: origin and topographic organisation. *Journal of Comparative Neurology* **236**, 161-178.

Armengol J.A., Sotelo C., Angaut P. & Alvarado-Mallart R.M. (1989). Organisation of host afferents to cerebellar grafts implanted into kainate lesioned cerebellum in adult rats. Hodological evidence for the specificity of host-graft interactions. *European Journal of Neuroscience* **1**,75-93.

Azizi S.A. & Woodward D.J. (1987). Inferior olivary nuclear complex of the rat. Morphology, and comments on the principle of organisation within the olivocerebellar system. *Journal of Comparative Neurology* **263**, 467-484.

Bähr M. & Eschweiler G.W. (1991). Regenerating adult rat retinal axons reconnect with target neurones *in vitro*. *Neuroreport* **2**, 581-584.

Bähr M. & Schwab M.E. (1996). Antibody that neutralises myelin-associated inhibitors of axon growth does not interfere with recognition of target-specific guidance information by rat retinal axons. *Journal of Neurobiology* **30**, 281-292.

Bähr M. & Wizenmann A.J. (1996). Retinal ganglion cell axons recognise specific cues present in the deafferented adult rat superior colliculus. *Journal of Neuroscience* **16**, 5106-5116.

Benedetti F. Montarolo P.G., Strata P. & Tosi L. (1983). Collateral reinnervation in the olivocerebellar pathway in the rat. In: B. Haber, J.R. Perez-Polo, G. Hashim and A.M.L. Giuffrida Stella (Eds.), *Nervous System Regeneration*, Vol 19, Alan Liss, new York, pp. 461-464.

Bravin M., Morando L., Vercelli A., Rossi F. & Strata P. (1996). Block of electrical activity leads to an increased spine number in Purkinje cell dendrites of adult rats. *European Journal of Neuroscience supplement.* **9** abstr. 20.15.

Bravin M., Savio T., Strata P. & Rossi F. (1997). Olivocerebellar axon regeneration and target reinnervation following dissociated Schwann cell grafts in surgically injured cerebella of adult rats. *European Journal of Neuroscience* **9**,2634-2649.

Brochu G., Maler L. & Hawkes R. (1990). Zebrin II: a polypeptide antigen expressed selectively by Purkinje cells reveals compartments in rat and fish cerebellum. *Journal of Comparative Neurology* **291**, 538-552.

Buffo A., Fronte M., Oestreicher B. & Rossi F. (1998). Degenerative phenomena and reactive modifications of the adult rat inferior olivary neurones following axotomy and disconnection from their target. *Neuroscience* **85**, 587-604.

Caroni P. & Schwab M.E. (1988a). Two membrane protein fractions from rat central myelin with inhibitory properties for neurite growth and fibroblast spreading. *Journal of Cell Biology* **106**, 1281-1288.

Caroni P. & Schwab M.E. (1988b). Antibody against myelin-associated inhibitor of neurite growth neutralises non permissive substrate properties of CNS white matter. *Neuron* **1**, 85-96.

Chédotal A., Bloch-Gallego E. & Sotelo C. (1996). The embryonic cerebellum contains topographic cues that guide developing inferior olivary axons. *Development* **124**, 861-870.

Chédotal A. & Sotelo C. (1992). Early development of olivocerebellar projection in the foetal rat using CGRP immunocytochemistry. *European Journal of Neuroscience* **4**, 1159-1179.

Fawcett J.W. (1992). Intrinsic neuronal determinants of regeneration. *Trends in Neuroscience* **15**, 5-8.

Gravel C., Eisenman L.M., Sasseville R. & Hawkes R. (1987). Parasagittal organisation of the rat cerebellar cortex: direct correlation between antigenic Purkinje cell bands revealed by mabQ113 and the organisation of the olivocerebellar projection. *Journal of Comparative Neurology* **265**, 295-310.

Guénard V., Xu X.M. & Bunge M.B. (1993). The use of Schwann cell transplantation to foster central nervous system repair. *Seminars in Neuroscience* **5**, 401-411.

Hankin M. & Lund R. (1991). How do retinal axons find their targets in the developing brain? *Trends in Neuroscience* **14**, 224-228.

Harvey A.R., Gan S.K. & Pauken J.M. (1987). Foetal tectum or cortical tissue transplanted into brachial lesion cavities in rats: influence on the regrowth of host retinal axons. *Journal of Comparative Neurology* **263**, 126-136.

Hawkes R. Colonnier M. & Leclerc N. (1985). Monoclonal antibodies reveal sagittal banding in the rodent cerebellar cortex. *Brain Research* **333**, 359-365.

Hawkes R. & Herrup K. (1995). Aldolase C/zebrin II and the regionalisation of the cerebellum. *Journal of Molecular Neuroscience* **6**, 147-158.

Herrup K. & Kuemerle B. (1997). The compartimentalisation of the cerebellum. *Annual Review of Neuroscience* **20**, 61-90.

Jacobson M.J. (1991). Developmental Neurobiology. London, New York: Plenum Press.

Keirstead S.A., Rasminsky M., Fukuda Y., Carter D.A., Aguayo A.J. & Vidal-Sanz M. (1989). Electrophysiologic response in hamster superior colliculus evoked by regenerating retinal axons. *Science* **246**, 255-256.

Kruger L., Benotti C., Rivolta R. & Samanin R. (1993). Distribution of GAP-43 in the adult rat brain. *Journal of Comparative Neurology* **333**, 417-434.

Kubota Y., Inagaki S., Shimada S., Kito S., Zaidi M., Girgis S.I., MacIntyre I. & Tohyama M. (1987). Transient appearance of the calcitonin gene-related immunoreactive fibres in the developing cerebellum of the rat. *Brain Research* **425**, 385-388.

Laskowski M. B. & Sanes J. R. (1988). Topographically selective reinnervation of adult mammalian skeletal muscles. *Journal of Neuroscience* **8**:3094-3099.

Morara S., Provini S. & Rosina A. (1989). CGRP expression in the rat olivocerebellar system during postnatal development. *Brain Research* **504**, 315-319.

Njå A. & Purves D. (1977). Reinnervation of guinea pig superior cervical ganglion cells by preganglionic fibres arising from different levels of the spinal cord. *Journal of Physiology (London)* **272**, 633-651.

Palay S.L. & Chan-Palay V. (1974). Cerebellar cortex. Cytology and organisation, Springer Berlin.

Rankin E.C. & Cook J.E. (1986). Topographic refinement of the regenerating retinotectal projection of the goldfish in standard laboratory conditions: a quantitative WGA-HRP study. *Experimental Brain Research* **63**, 409-420.

Rossi F, Fronte M., Hawkes R. & Strata P. (1997). Innervation specificity of transplanted embryonic Purkinje cells by adult climbing fibres. *Society for Neuroscience Abstracts* 656.2.

Rossi F., Jankovski A. & Sotelo C, (1995). Differential regenerative response of Purkinje cell and inferior olivary axons confronted with embryonic grafts: environmental cues versus intrinsic neuronal determinants. *Journal of Comparative Neurology* **359**:663-677.

Rossi F. & Strata P. (1995). Reciprocal trophic interactions in the adult climbing fibre-Purkinje cell system. *Progress in. Neurobiology* **47**, 341-369.

Rossi F., Wiklund L., van der Want J.J.L. & Strata P. (1991a). Reinnervation of cerebellar Purkinje cells by climbing fibres surviving a subtotal lesion of the inferior olive in the adult rat. I Development of new collateral branches and terminal plexuses. *Journal of Comparative Neurology* **308**, 513-535.

Rossi F. , Wiklund L., van der Want J.J.L. & Strata P. (1991b). Reinnervation of cerebellar Purkinje cells by climbing fibres surviving a subtotal lesion of the inferior olive in the adult rat. II Synaptic organisation on reinnervated Purkinje cells. *Journal of Comparative Neurology* **308**, 536-554.

Rouse R.V. & Sotelo C. (1990). Grafts of dissociated cerebellar cells containing Purkinje cell precursors organise into zebrin I defined compartments. *Experimental Brain Research* **82**, 401-407.

Scheibel M.E. & Scheibel A.B. (1954). Observations of the intracortical relations of climbing fibres of the cerebellum. *Journal of Comparative Neurology* **101**, 733-763.

Schwab M.E., Kapfhammer J.P. & Bandtlow C.E. (1993). Inhibitors of neurite growth. *Annual Review of Neuroscience* **16**, 565-595.

Sotelo C., Bourrat F. & Triller A. (1984). Postnatal development of the inferior olivary complex in the rat: II. Topographic organisation of the immature olivocerebellar projection. *Journal of Comparative Neurology* **222**, 177-199.

Sperry R.W. (1963). Chemioaffinity in the orderly growth of nerve fibre pattern and connections. *Proceedings of the National Academy of Science USA* **50**, 703-710.

Stuermer C.A.O. (1988). Trajectories of regenerating retinal axons in the goldfish tectum: I. A comparison of normal and regenerated axons at late regeneration stages. *Journal of Comparative Neurology* **267**, 55-68.

Thanos S., Naskar R. & Heiduschka P. (1997). Regeneration of ganglion cell axons in the adult rat establishes retinofugal topography and restore visual function. *Experimental Brain Research* **11**, 483-491.

Wassef M., Angaut P., Arsenio-Nunes L., Bourrat F. & Sotelo C. (1992). Purkinje cell heterogeneity: its role in organising the cerebellar cortex connections. In: R. Llinás and C. Sotelo (Eds.), *The cerebellum Revised*, Springer-Verlag, New York, pp. 5-21.

Wizenmann A.E., Thies S., Klostermann F., Bonhoeffer F. & Bähr M. (1993). Appearance of target-specific guidance information for regenerating axons after CNS lesions. *Neuron* **11**, 975-983.

Zagrebelsky M., Rossi F., Hawkes R. & Strata P. (1996). Topographically organised climbing fibre sprouting in the adult rat cerebellum. *European Journal of Neuroscience* **8**, 1051-1054.

Zagrebelsky M., Strata P., Hawkes R. & Rossi F. (1997). Reestablishment of the olivocerebellar projection map by compensatory transcommissural reinnervation following unilateral transection of the inferior cerebellar peduncle in the newborn rat. *Journal of Comparative Neurology* **379**, 283-299.

Zwimpfer T.J., Aguayo A.J. & Bray G.M. (1992) Synapse formation and preferential distribution in the granule cell layer by regenerating retinal ganglion cfell axons guided to the cerebellum of adult hamsters. *Journal of Neuroscience* **12**,1144-1159.

Part 3

Models and Theoretical Approaches

Regulatory effects of long term biochemical processes in integrate-and-fire model neurons

Marco Bove, Michele Giugliano, Massimo Grattarola

Bioelectronics and Neurobioengineering Group
Department of Biophysical and Electronic Engineering, University of Genoa
V. l'Opera Pia 11A, I-16145 Genova, Italy

Keywords. Model neurons, Dynamic regulation, Integrate-and-fire

1 Introduction

The understanding of the dynamics of networks of biological neurons, by means of mathematical descriptions, computer simulations (20) and artificial emulations of their electrophysiological properties (7, 8, 15, 16), is a central issue in computational neuroscience research. Many models have been proposed and a broad class of them shares the same qualitative features (1, 5).
We can outline two basic approaches: models that account for accurate sub-cellular biophysical phenomena (5, 10, 13, 17), and models that provide input-output mapping, stating *a priori* the role of signals under a computational perspective (1, 9, 11, 12, 20).
The first group is related mainly to the mathematical description of ionic flows through a patch of a lipidic bilayer, i.e. a plasmatic membrane, originally proposed by Hodgkin and Huxley (10). This approach, further improved after recent outstanding progresses in electrophysiological measurement techniques and the availability of advanced experimental neurobiological data (4, 6, 22), is characterised by a common repertoire of oscillatory/excitable processes and by the definition of the non-linear ionic membrane permeability.
A quite different approach to model the nervous system and emergent collective behaviours of large assemblies of functionally connected neurons consists in neglecting much of the biological complications and simplifying dynamic properties of the single units, in order to gain some insight on the characteristics of the whole system (1, 13). Even if the second modelling approach had a strong impact in development of the theory of formal neural computation and statistical learning theory (24), in particular clarifying the huge knowledge representation power of non-linear adaptive base functions (9, 20, 24), the search for intermediate mathematical description and the investigation on properties of more realistic model neuronal networks seem nowadays more interesting under the point of view

of neuroscience basic research (1, 2, 13) and of artificial neural networks - neuromorphic electronic systems design (7, 15, 16).
In this work we study second-order biochemical phenomena, not easily tractable with complex biophysical models when dealing with wide-scale networks of model neurons. In section 3.1, a mathematically consistent procedure of reduction of a conductance-based model neuron is used to define intrinsic electrical parameters of the integrate-and-fire model with refractoriness (1). Electrical properties of the consistently determined simplified model are used to reproduce the same qualitative, and in some extent quantitative, properties of the complex model neuron investigated in section 2.
A simple model for intracellular calcium concentration dynamics (25) is reviewed in section 2.2, and its reduced description is formalised in section 3.2, in order to introduce such a chemical messenger system in the integrate-and-fire model neuron. In section 3.2 the attempt at introducing slow second-order phenomena related to the channel synthesis and removal into the simplified model will be proved to be successful, maintaining the stabilising features of the regulatory mechanisms, investigated in section 2.3 in the case of full conductance-based model neuron (2).

2 Biophysical Modelling

2.1. Conductance-based Model Neuron

Focusing our interest on single-compartment modelling (18, 19), i.e. the space-clamp hypothesis assumed for each point of the excitable membrane, we introduce the deterministic formalism (6, 10, 22) for the temporal dynamics of the membrane voltage V(t) [mV]. Assuming almost perfect dielectric properties for the lipidic bilayer that constitutes the membrane, equation [1] follows from charge conservation considerations.

[1] $$C_m \cdot \frac{dV}{dt} = I_{ext} - f(V) \qquad I_{ext} = I_{stim} + I_{syn}$$

In the previous equation, we denoted with C_m the equivalent electric capacitance per area unit [μF/cm^2], with f(V) the voltage-dependent current densities [μA/cm^2] related to the ionic non-linear permeability features of the membrane, and with I_{ext} the total contribution to external current density [μA/cm^2], including external stimulus currents I_{stim} and metabolic contributions from active transport mechanisms, i.e. ionic pumps. Since metabolic contributions are small, we consider them negligible in the foregoing description, assuming thus time-independent the equilibrium potentials for potassium and calcium ions, E_K and E_{Ca} respectively.

For the sake of simplicity, we consider the model proposed by Morris and Lecar (17) which provides a general reduction of complexity, as compared to more detailed model neurons (3, 5, 10), and although it was originally developed as a mathematical description of the barnacle giant muscle fibre, it proves to be suited for modelling of excitatory properties of other cellular systems (17, 18).
The model is based on two non-inactivating conductances for K^+ and Ca^{2+} ions, uniformly distributed on the surface of the membrane. The expression of the total ionic current density f(V) can be made explicit:

[2] $$f(V) = \left(I_{leak} + I_{Ca} + I_K\right)$$

[3] $$I_{leak} = \bar{g}_{leak} \cdot \left(V - E_{leak}\right)$$

[4] $$I_{Ca} = \bar{g}_{Ca} \cdot m_\infty \cdot \left(V - E_{Ca}\right) \qquad m_\infty = \frac{1}{2} \cdot \left[1 + \tanh\left(\frac{V - V_1}{V_2}\right)\right]$$

[5] $$I_K = \bar{g}_K \cdot n \cdot \left(V - E_K\right)$$

Substituting equation [2] in equation [1], the final form of the differential equation is derived, accounting integrative-capacitive properties of the plasmatic membrane.

[6] $$C_m \cdot \frac{dV}{dt} = I_{ext} - \left[\bar{g}_{leak} \cdot \left(V - E_{leak}\right) + \bar{g}_{Ca} \cdot m_\infty \cdot \left(V - E_{Ca}\right) + \bar{g}_K \cdot n \cdot \left(V - E_K\right)\right]$$

The description of the model is completed by the differential equation satisfied by the state variable n, representing the fraction of open K^+ protein channels.

[7] $$\frac{dn}{dt} = \left[\frac{1}{\tau_n} \cdot \cosh\left(\frac{V - V_3}{2 \cdot V_4}\right)\right] \cdot \left(n_\infty - n\right) \qquad n_\infty = \frac{1}{2} \cdot \left[1 + \tanh\left(\frac{V - V_3}{V_4}\right)\right]$$

Equation [7] expresses the non-linear voltage dependence of molecular structures determining the functional state of protein channel, accounting for the generation of action potentials and for the membrane refractoriness.
The general analysis of the temporal evolution of state variables V and n, under different stimulus conditions and using different sets of parameters, is similar to those studied in the case of other biophysical model neurons, retaining the Hodgkin-Huxley format (3, 10). Small amplitude external current stimuli produce passive membrane depolarisation, unless a threshold voltage is reached, when oscillations in V(t) occur (Fig. 3.1, left).

2.2 Intracellular Ca^{2+} Concentration Modelling

The effect of active ionic pumps and the release of free Ca^{2+} ions from intracellular stores are neglected and Ca^{2+} buffering is described by an exponential decay, i.e. a spontaneous decrease of the intracellular levels. This is characterised by a decay rate, proportional to actual amount of Ca^{2+} ions according to a time-invariant first order kinetic scheme. Any increase in the intracellular calcium concentration is due to the inward calcium currents only. This contribution is sensible especially during action potentials, when most of Ca^{2+} channels open.
The model we derive is far from a detailed description of all biochemical pathways involved in determining the actual Ca^{2+} concentration in anatomical structures of a real cell. Nonetheless, this simple model is accurate enough to make explicit the existing feedback of the electrical activity on the dynamics of a variety of sub-cellular chemical signalling systems, via calcium-mediated single or second-messenger elements.
Let us consider a generic closed regular surface S of total area Σ [cm^2] and enclosed volume Ω [cm^3] and characterised by Ca^{2+} permeability properties, uniformly distributed on the surface. Moreover let us consider these properties to be continuos (i.e. the discrete nature of the ionic flows through a biological membrane is not being considered). Regarding the total number of free Ca^{2+} ions inside the surface S as N(t), it follows by definition:

[8] $$N(t) = \{[Ca^{2+}]_{IN}(t) \cdot \Omega\} \qquad [Ca^{2+}]_{IN}: \frac{\#\text{ions}}{cm^3}$$

We assume that the space gradient $\overline{\nabla}[Ca^{2+}]_{IN} = \overline{0}$ for each (x,y,z) inside the volume, so that any functional dependence on spatial variables can be neglected. We express the concentration as the number of particles per volume unity, or alternatively in terms of μM by multiplying $[Ca^{2+}]_{IN}$ times the conversion factor equal to $(6.02\ 10^4)^{-1}$ μM cm^3.
As previously noted the increase of the number of Ca^{2+} ions during a very small time interval Δt can be related to the difference between an inward and an outward flow:

[9] $$\{[Ca^{2+}]_{IN}(t+\Delta t) - [Ca^{2+}]_{IN}(t)\} \cdot \Omega = \Phi_{IN}(t) \cdot \Delta t - \Phi_{OUT}(t) \cdot \Delta t$$

$\Phi_{OUT} = \xi \cdot [Ca^{2+}]_{IN} \cdot \Omega$ from the action mass law, ξ^{-1} being the time constant of spontaneous decaying. In the case of the Morris-Lecar model neuron, we can explicitly define the inward flow rate for Ca^{2+} ions, Φ_{IN}. Since we have defined the total deterministic calcium current crossing the membrane (inward current) as

[10] $$(-I_{Ca}) = g_{Ca}(t) \cdot (E_{Ca} - V)$$

from the previous description, we can state that the inward calcium flow has the following expression:

$$[11] \qquad \Phi_{IN} \cdot \Delta t = \iint_S \frac{(-I_{Ca}) \cdot dS}{z_{Ca} \cdot q} \cdot \Delta t$$

with $z_{Ca} = 2$ is the valence of Ca^{2+} ions and $q \approx 1.6\ 10^{-19}$ C the elementary charge. Implicitly assuming the ergodicity of stochastic processes associated with opening and closing of the functional gating mechanisms of a single protein channel, we can conclude that terms under the integral operator in equation [11], do not depend on the position on the surface S. The following differential equation can be derived when $\Delta t \rightarrow 0$:

$$[12] \qquad \frac{d[Ca^{2+}]_{IN}}{dt} = -\xi \cdot \left(A \cdot I_{Ca} + [Ca^{2+}]_{IN}\right)$$

We expressed $[Ca^{2+}]_{IN}$, the intracellular calcium concentration, in μM. As a consequence, A is a geometric factor whose explicit expression is reported below:

$$[13] \qquad A = \frac{\Sigma}{\Omega} \cdot \frac{(6.02 \cdot 10^{14})^{-1}}{z_{Ca} \cdot q \cdot \xi} \frac{\mu M \cdot ms \cdot cm^2}{C}$$

and it can be proved that in the case of a spherical surface S, A is proportional to the inverse of the radius of the sphere, acting as a scaling factor with regards to the effective contribution in the increase of intracellular calcium concentration due to the calcium current I_{Ca}. For the sake of simplicity, in the following we express $[Ca^{2+}]_{IN}$ in A - units, i.e. A is assumed equal to 1. Simulation results of eqs. [6], [7] and [12] are reported in Fig. 3.3, upper panel.

2.3 Protein Synthesis Modelling

Recently, a new view to look at traditional biophysical model was proposed in literature (2, 14, 21). Mathematical descriptions of an excitable membrane usually regard its dynamics to be time-invariant. This assumption holds over a faster time scale than those associated with slow ATP-dependent chemical machinery and metabolic changes, determining electrical properties of neurons.

For instance, over long time scales the number and the characteristics of protein structures that allow ionic permeability through the membrane are not constant quantities. Biochemical processes such as protein phosphorylation and gene expression, can modulate the density of channels in the membrane, regulating

inward and outward currents, as already observed in experimental studies on cultured neurons (23).
This is particularly meaningful when noting that the effective life-time of a nervous cell is comparable with that of the animal it was explanted from, and the electrical properties remain fixed and stable over a long time, under metabolic activities and other ageing processes, acting at the molecular scale. Actually, channel densities and thus the maximal conductances $\bar{g}_i$ control membrane voltage dynamics, affecting the existence, the shape and the frequency of any oscillatory evolution V(t) (see Fig. 2.1).
Following Abbott et al. (2, 14) we introduce a slow process affecting the value of the maximal conductances $\bar{g}_i$, modelling the insertion and degradation of channels in the membrane. We can relate the change in the total number of ionic channels to the maximal ionic conductance of a biophysical model, expressing explicitly channel synthesis and removal rates:

$$[14] \qquad \frac{d\bar{g}_i}{dt} \propto S_i - R_i$$

If we assume that such process could be linked to electrical activity of the cell, using the intracellular calcium concentration as a feedback element, highly correlated with instantaneous firing rate, the following regulatory differential equations can be formalised:

$$[15] \qquad \tau_{reg} \cdot \frac{d\bar{g}_{Ca}}{dt} = G_{Ca} \cdot \sigma\left(\frac{C_T - [Ca^{2+}]_{IN}}{\Delta}\right) - \bar{g}_{Ca}$$

$$[16] \qquad \tau_{reg} \cdot \frac{d\bar{g}_K}{dt} = G_K \cdot \sigma\left(\frac{[Ca^{2+}]_{IN} - C_T}{\Delta}\right) - \bar{g}_K$$

$\sigma(x) = 1 / (1+\exp(-x))$ is the standard sigmoidal function and it links the channel synthesis driving force to the displacement of the intracellular calcium concentration from an ideal target value C_T. Δ specifies the maximum slope of the sigmoid, G_K and G_{Ca} the maximum value, under regime conditions, for $\bar{g}_{Ca}$ and $\bar{g}_K$ respectively and finally τ_{REG} represents the time constant of the spontaneous exponential decaying of the number of channels, modelling the protein removal rate. The last is assumed to be activity-independent. In the model neuron, a specific mechanisms to achieve, restore and maintain stable firing pattern over long time period is provided by a dynamic regulation of the intrinsic electrical properties of the membrane, opposed to channels turnover, cell growth and variable external environmental conditions (2). The full regulated model (eqs. [6],[7], [12], [15] and [16]) reveals interesting behaviours: the regulatory mechanism spontaneously synthesise the current densities needed to achieve a specific pattern of electrical activity, from a wide variety of initial values for the maximal conductances. Even starting with no protein channels, under constant current stimulation, the two maximal conductances approach stable regimes values, as the state-plane trajectory reveals (Fig. 2.1).

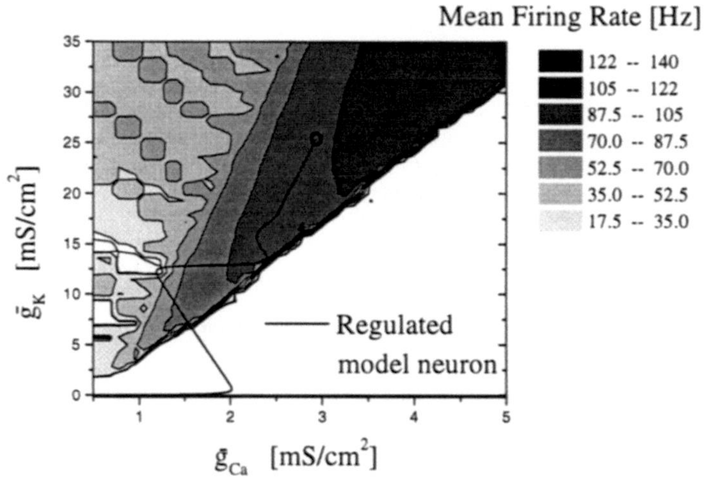

Fig. 2.1. The maximal-conductance plane for the Morris-Lecar model neuron. Under a constant external current stimulation, different mean firing rate regimes are described for each point of the plane by grey levels. The trajectory superimposed to the peninsula of oscillatory behaviour, is induced by the slow regulatory processes, starting with no ionic channels interleaved in the membrane (i.e. initial conditions are set to the origin of the plane). The thick circle represent the stable equilibrium point.

Moreover, significant stability enhancement to environmental perturbation occurs, as clearly showed in Fig. 2.2.

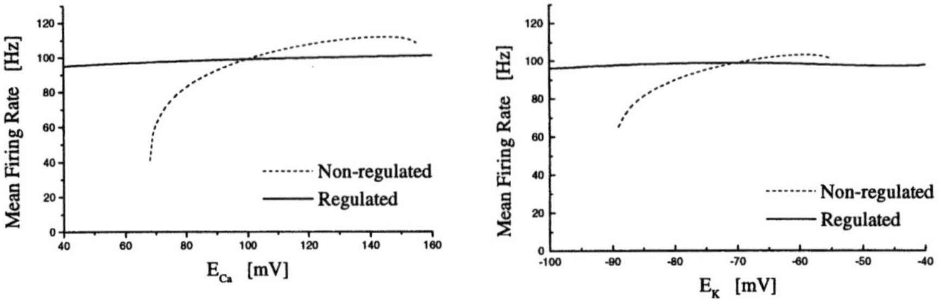

Fig. 2.2. Robustness enhancement in the Morris-Lecar regulated model neuron to environmental perturbations: slow changes in ionic equilibrium potentials greatly affect the mean firing rate. On the contrary, dynamic regulation stabilises the intracellular calcium concentration thus achieving a constant stable firing pattern.

3 The Integrate-and-fire Model Neuron

3.1 Reduction of the Morris-Lecar Model Neuron

As we already discussed in section 1, choosing the accuracy of a mathematical description in order to gain some insights on the essence of a natural phenomena is a challenging problem in computational neuroscience. The choice of conductance-based models or other simpler descriptions depends on the degree of abstraction required for the current investigation, even if mathematically and conceptually simpler models often lack of any direct link to the properties of more complex ones.

Every reduction introduces assumptions never verified in real experiments under the framework that fine details are unnecessary to capture their basic essence. But it is important to retain as many correlations as possible with regards to experimentally measured properties in order to perform quantitative comparisons between theoretical predictions and real measures.

For instance, any reduction of the unregulated Morris-Lecar model neuron (eqs. [6], [7]) corresponds to ignore one of the two equations. We are interested in preserving integrative-capacitive electrical properties of nervous cells and we assume that action potential generation and refractoriness mechanisms are to be regarded as highly stereotyped dynamic behaviours. From these considerations, we can neglect the equation [7] reducing the accuracy in describing biophysical activation variables. In other words, because of the invariance in the shape of action potentials, specifying repolarisation, hyperpolarisation and the refractory period, the precise shaping of the action potential is not relevant in describing computational properties and collective auto-organising properties of a network.

The same set of phenomena can be efficiently described with a lower degree of fidelity and considerably reducing the times of computer simulations.
Let us consider the dynamics of n to be instantaneous, compared with the passive time constant associated with equation [1]. We can rewrite equation [1] as

$$[17] \qquad C_m \cdot \frac{dV}{dt} \cong I_{ext} - h(V)$$

$$h(V) = \bar{g}_{leak} \cdot (V - E_{leak}) + \bar{g}_{Ca} \cdot m_\infty \cdot (V - E_{Ca}) + \bar{g}_K \cdot n_\infty \cdot (V - E_K)$$

The previous equation constitutes a good approximation for the description of the membrane voltage, under the excitability threshold, when V is close to its resting value. By definition, $I_{ext} = 0\ \mu A/cm^2$ implies that the second term of the previous equation is close to zero, where $V = V_o$, where V_o is the resting membrane potential. Since this implies $dV/dt = 0$, we can provide a Taylor series approximation of the equation [17], whose leading term is a reduced linear description of the subthreshold dynamics.

$$[18] \qquad C_m \cdot \frac{dV}{dt} \cong I_{ext} + \bar{g} \cdot (V_o - V) \qquad V < V_{th}$$

Considering action potentials as stereotyped behaviours we need to introduce a threshold value V_{th}, so that once V reaches V_{th} it is reset back to an hyperpolarising voltage V_{RESET} and it is refractory to any external perturbation during a subsequent fixed time interval τ_{REF}. We determined an equivalent integrate-and-fire model whose complete mathematical formalisation follows:

$$[19] \qquad \begin{cases} C_m \cdot \dfrac{dV}{dt} = I_{ext} + \bar{g} \cdot (V_o - V) & V < V_{th} \\ V(t) = V_{RESET} & t - t_k < \tau_{REF} \quad V(t_k^-) = V_{th} \\ V(t_k^+) = V_{MAX} & \end{cases}$$

C_m, $\bar{g}$, V_o are directly linked to the original passive model properties, V_{RESET} and V_{MAX} are chosen by computing the lower and higher bound to the dynamics of V(t) in the Morris-Lecar model, while V_{th} and τ_{REF} were fixed with an optimising algorithm, based on a gradient descent algorithm, since the mean firing rate can be expressed analytically:

$$[20] \qquad \nu(V_{th}, V_{RESET}, \tau_{REF}) = \left\{ \tau_{REF} + \frac{C_m}{\bar{g}} \cdot \ln\left[\frac{V_{RESET} - (V_o + I_{stim}/\bar{g})}{V_{th} - (V_o + I_{stim}/\bar{g})} \right] \right\}^{-1}$$

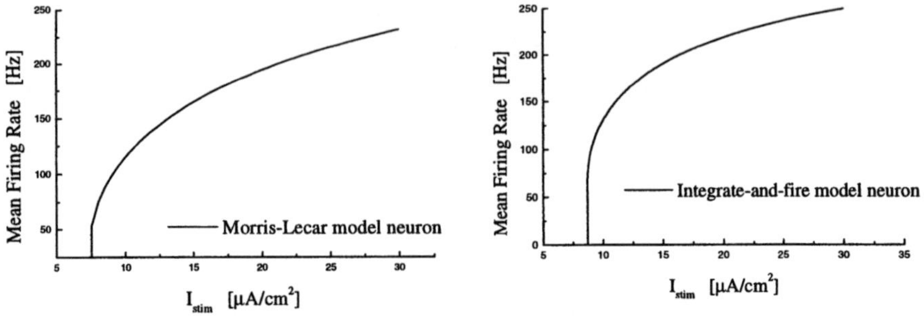

Fig. 3.1. The dependence of the mean firing rate on the amplitude of constant current stimulation, for (left) the Morris-Lecar unregulated model neuron and (right) the Integrate-and-Fire unregulated model neuron.

The comparison between the temporal evolution of membrane voltage of the Morris-Lecar model neuron and those of the integrate-and-fire model neuron, is reported in Fig 3.2, where a constant current stimulus $I_{ext} = I_{stim}$ is provided during a limited temporal window.

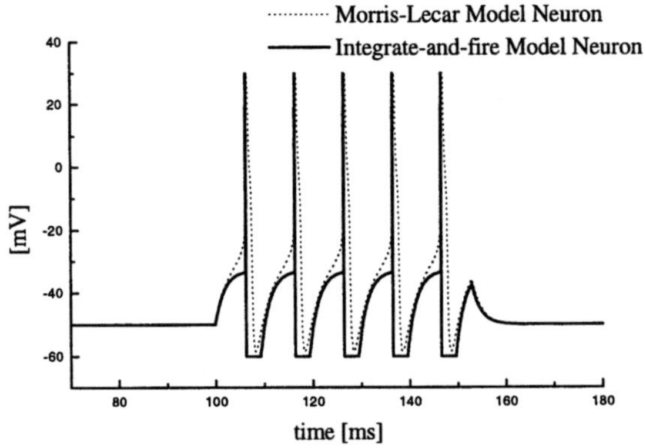

Fig. 3.2. The temporal evolution of the membrane potential, under a constant current pulse stimulation (between 100 and 152 ms) is compared between Morris-Lecar and integrate-and-fire model neurons

3.2 Reduced Calcium Model

In the previous paragraphs, we reviewed the voltage dependence for the deterministic fraction of channels, selective to Ca^{2+} ions, in the open state. The main hypothesis underlying the mathematical reduction of the unregulated Morris-Lecar model neuron consists in the following statement:

$V(t) \cong V_o$ for most of the time.

With the numerical parameters chosen for the simulations reported, it follows immediately that

$V(t) << (V_1 - V_2)$

and then $m_\infty(t) \cong 0$ for most of the time. We could therefore state that for most of the time

$$[21] \qquad \frac{d[Ca^{2+}]_{IN}}{dt} \approx -\xi \cdot \left([Ca^{2+}]_{IN} + 0\right)$$

We now can recover the dynamics of the inward calcium current, by stereotyping the involved processes. First of all, we observe that during action potentials:

$V(t) >> (V_1 + V_2)$ and then $m_\infty \cong 1$, $\left(-I_{Ca}\right) \cong \bar{g}_{Ca} \cdot \left(E_{Ca} - V\right)$

For the sake of simplicity, we assume that the duration of action potentials has a fixed value τ_{AP} [ms] (lesser or equal to the refractory period) for every external current. We can further approximate that V(t) >> 0 mV during the full duration of action potential, while V(t) << 0 mV for the rest of the time.
From these considerations, we approximate the inward flow rate of Ca^{2+} ions as:

$$[22] \qquad \Phi = \sum_i Q \cdot \Pi\left(\frac{t - t_i}{\tau_{AP}}\right) \qquad \Pi(x) = \begin{cases} 1 & x \in (0;1) \\ 0 & \text{else} \end{cases}$$

In equation [22], we denoted with t_1, t_2, t_3, ... the times of occurrence of action potentials, i.e. threshold crossing times in the integrate-and-fire model neuron.
From equations [21] and [22], we can express the final reduced model for calcium concentration:

$$[23] \qquad \frac{d[Ca^{2+}]_{IN}}{dt} \approx \begin{cases} -\xi \cdot \left([Ca^{2+}]_{IN} + Q\right) & \left(t - t_i\right) < \tau_{AP} \\ -\xi \cdot \left([Ca^{2+}]_{IN} + 0\right) & \text{else} \end{cases}$$

Equation [23] is a linear, time-invariant, differential equation that can be solved in closed form and calculated in a computationally not intensive way.

$$[Ca^{2+}]_{IN}(t) = \begin{cases} [Ca^{2+}]_{IN}(t_i) \cdot e^{-\xi(t-t_i)} & (t-t_i) > \tau_{AP} \;\; t < t_{i+1} \\ [Ca^{2+}]_{IN}(t_{i+1}) \cdot e^{-\xi(t-t_{i+1})} + Q \cdot \left(1 - e^{-\xi(t-t_{i+1})}\right) & (t-t_{i+1}) < \tau_{AP} \end{cases}$$

[24]

Let us consider a constant time bin Δt, and a uniform discretisation of time axis. We can iteratively compute the instantaneous value for calcium concentration at the next time step, as follows:

$$[Ca^{2+}]_{IN}((k+1)\cdot\Delta t) = \begin{cases} [Ca^{2+}]_{IN}(k\cdot\Delta t)\cdot e^{-\xi\cdot\Delta t} & \begin{cases}(k\cdot\Delta t - t_i) > \tau_{AP} \\ (k+1)\cdot\Delta t < t_{i+1}\end{cases} \\ [Ca^{2+}]_{IN}(k\cdot\Delta t)\cdot e^{-\xi\cdot\Delta t} + Q\cdot(1-e^{-\xi\cdot\Delta t}) & (k\cdot\Delta t - t_{i+1}) < \tau_{AP} \end{cases}$$

[25]

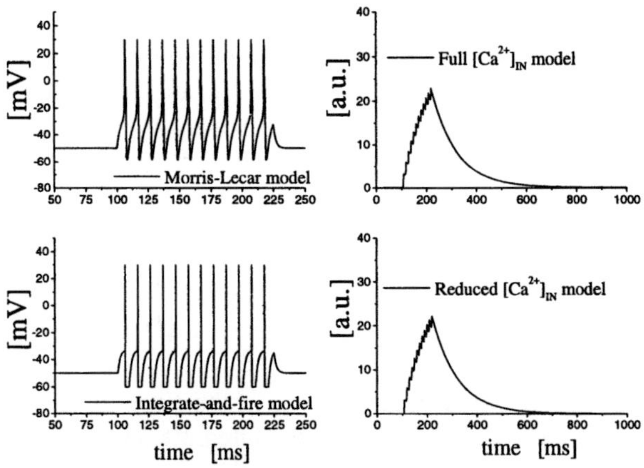

Fig. 3.3. Temporal evolution of **(left)** V(t) [mV] and **(right)** $[Ca^{2+}]_{IN}$ [a.u.], under constant current pulse stimulation in **(upper)** Morris-Lecar and **(lower)** integrate-and-fire model neurons.

3.3 Regulated Integrate-and-fire Model Neuron

In order to implement dynamic regulation into the integrate-and-fire model neuron, as described in section 2.3, a convenient reduced model for calcium ion dynamics has been provided. Since the good accuracy achieved in approximating equation [12] with equation [23] (see Fig. 3.3, right side), it is possible to define a regulatory equation for the equivalent maximal conductance in the integrate-and-fire model neuron. We propose to use precisely the same format of equations [15] and [16]:

$$[26] \qquad \tau_{ref} \cdot \frac{d\bar{g}}{dt} = G \cdot \sigma\left(\frac{[Ca^{2+}]_{IN} - C_T}{\Delta}\right) - \bar{g}$$

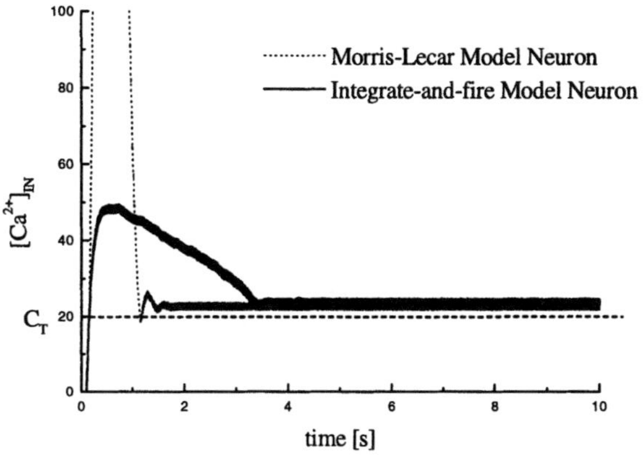

Fig. 3.4. Temporal evolution of the intracellular calcium concentration in the regulated integrate-and-fire model neuron, compared with the regulated Morris-Lecar model neuron, under a constant current stimulation I_{stim}.

Simulations of the full regulated integrate-and-fire model neuron (eqs. [19], [23], [26]) are reported in Fig. 3.4, 3.5 and 3.6. An asymptotically stable behaviour in the conductance space, whose dimension is 1, is preserved together with the exponential evolution with time constant τ_{REG} for the maximal conductance $\bar{g}$ (Fig. 3.6). Calcium concentration follows qualitatively the same C_T - tracking dynamics, shown by the regulated Morris-Lecar model neuron (Fig. 3.4), and a highly stabilising effect can be verified by comparing the dependence of the mean firing rate, under a constant current stimulation, on the V_{th} value, with the unregulated case. We chose V_{th} as an environmental perturbation in the reduced model (Fig. 3.5).

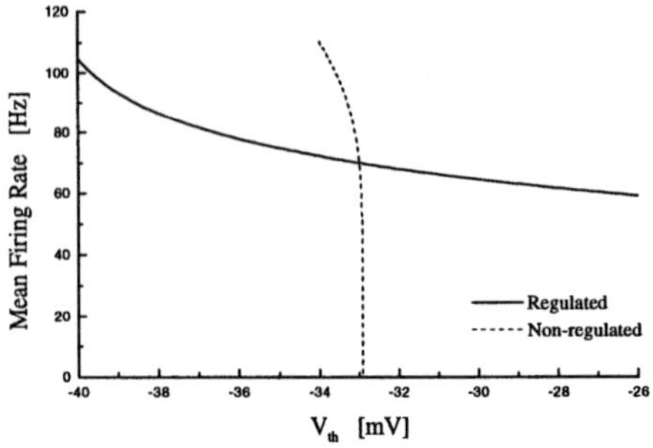

Fig. 3.5. Robustness enhancement in the regulated integrate-and-fire model neuron to slow changes of V_{th}, compared with those of the unregulated (G = 1 mS/cm^2).

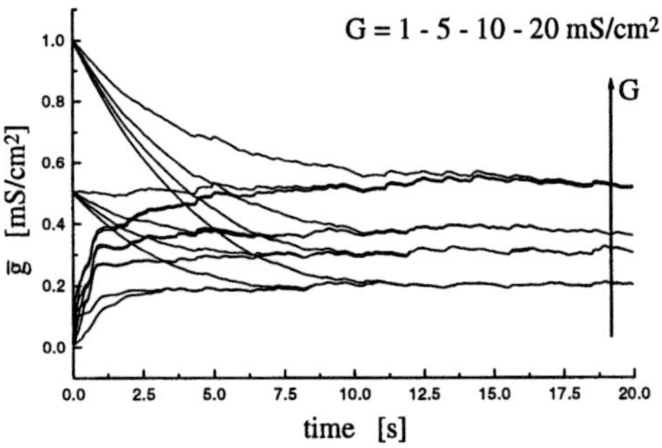

Fig. 3.6. A stable equilibrium value for the maximal conductance in the regulated integrate-and-fire model neuron is reached starting from a variety of initial conditions (0.01 - 0.1 - 0.5 - 1.0 mS/cm^2) , and it is a non-decreasing function of G.

4 Discussion

In the present work we examined the mathematical modelling of second-order slow biochemical pathways regulating current densities a conductance-based model neuron. The approach of modelling these slow adaptive and stabilising phenomena is computationally demanding when dealing with wide-scale networks of model neurons. We showed that the same regulatory mechanism could be directly introduced in a simplified description of neuronal excitability particularly well suited to be computer simulated in large assemblies of connected units.

A quantitative reduction procedure was used to derive the extended integrate-and-fire model neuron, including intracellular calcium dynamics. The reduced dynamically regulated model kept the same properties of the full model and it offered a new perspective for theoretical investigation of coherent emergent behaviours that characterise a network of neurons slowly adaptively synthetising channels and thus modulating their own intrinsic electrical properties.

Acknowledgments

This work was suported by University of Genoa (Italy).

References

1. Abbott, L.F., Kepler, T.B.: Model Neurons: from Hodgkin-Huxley to Hopfield. In: Garrido, L. (ed.): Statistical Mechanics of Neural Networks. Barcelona: Springer Verlag 1990, pp. 5-18
2. Abbott, L.F., LeMasson, G.: Analysis of Neuron Models with Dynamically Regulated Conductances. Neural Computation **5**, 823-842 (1993)
3. Buchholtz, F., Golowasch, J., Epstein, I.R., Marder, E.: Mathematical Model of an Identified Stomatogastric Ganglion Neuron. J. Neurophys. **67**, 332-349 (1992)
4. DeFelice, L.J., Clay, J.R.: Membrane Current and Membrane Potential form Single-Channle Kinetics. In: Sackmann, B., Neher, E. (eds): Single-Channel recording, 1983, cap 15.
5. Destexhe, A., Babloyantz, A., Sejnowski, T.J.: Ionic Mechanism for Intrinsic Slow Oscillations in Thalamic Relay Neurons. Biophys. J. **65**, 1538-1552 (1993)
6. Destexhe, A., Mainen, Z., Sejnowski, T.J.: Synthesis of Models for Excitable Membranes, Synaptic Transmission and Neuromodulation using a common Kinetic Formalism. J. Comp. Neurosci. **1**, 195-230 (1994)
7. Mahowald, M., Douglas, R.: A silicon neuron. Nature **354**, 515-518 (1991)
8. Grattarola, M., Bove, M., Martinoia, S., Massobrio, G.: Silicon Neuron simulation with SPICE: tool for neurobiology and neural networks. Med. & Bio. Eng. & Comp. **33**, 533-536 (1995)
9. Herz, J., Krogh, A., Palmer, R.G.: Introduction to the theory of neural computation. Lecture Notes Volume I, Santa Fe Institute. Addison-Wesley 1990
10. Hodgkin, A.L., Huxley, A.F.: A quantitative description of membrane current and its application to conduction and excitation of nerve. J. Physiol. **117**, 500-544 (1952)
11. Hopfield, J.J.: Neural Networks and physical systems with emergent collective computational abilities. Proc. Natl. Acad. Sci. USA **79**, 2554-2558 (1982)

12. Hopfield, J.J.: Neurons with graded response have collective computational properties like those of two-state neurons. Proc. Natl. Acad. Sci. USA **81**, 3088-3092 (1984)
13. Kepler, T.B., Abbott, L.F., Marder, E.:Reduction of conductance-based neuron models. Biol. Cybern. **66**, 381-387 (1992)
14. LeMasson, G., Marder, E., Abbott, L.F.: Activity-Dependent Regulation of Conductances in Model Neurons. Science **259**, 1915-1917 (1993)
15. Mead, C.: Neuromorphic Electronic Systems. Proc. IEEE **78**, 1629-1636 (1990)
16. Mead, C.: Analog VLSI Neural Systems. The Computation and Neural System Series. Addison-Wesley 1989
17. Morris, C., Lecar, H.: Voltage oscillations in the barnacle giant muscle fiber. Biophys. J. **35**, 193-213 (1981)
18. Rinzel, J., Ermentrout, G.B.: Analysis of Neural Excitability and Oscillations. In: Koch, C., Segev, I. (eds.): Methods in Neuronal Modelling: From Synapses to Networks. Mit Press 1989, pp. 135-169
19. Segev, I., Fleshman, J.W., Burke, R.E.: Compartmental Models of Complex Neurons. In: Koch, C., Segev, I. (eds.): Methods in Neuronal Modelling: From Synapses to Networks. Mit Press 1989, pp. 63-96
20. Sejnowski, T.J., Koch, C., Churchland, P.S.: Computational Neuroscience. Science **241**, 1299-1306 (1988)
21. Siegel, M., Marder, E., Abbott, L.F.: Activity-dependent current distributions in model neurons, Proc. Natl. Acad. Sci. USA **91**, 11308-11312 (1994)
22. Strassberg, A.F., DeFelice, L.J.: Limitations of the Hodgkin-Huxley Formalism: Effects of Single Channel Kinetics on Transmembrane Voltage Dynamics, Neural Comp. **5**, 843-855 (1993)
23. Turrigiano, G., LeMasson, G., Marder, E.: Selective Regulation of Current Densities Underlies Sponteneus Changes in the Activity of Cultured Neurons. J. of Neurosci. **15**, 3640-3653 (1995)
24. Vapnik, V.N.: The Nature of Statistical Learning. Springer 1995
25. Yamada, W.M., Koch, C., Adams, P.R.: Multiple Channels and Calcium Dynamics. In: Koch, C., Segev, I. (eds.): Methods in Neuronal Modelling: From Synapses to Networks. Mit Press 1989, pp. 97-133

A Formal Approach to the Translation of Cortical Maps

Klaas E. Stephan[1] and Rolf Kötter[1,2]

[1] C. & O. Vogt Institute for Brain Research, HHU Düsseldorf, D-40225 Düsseldorf, Germany
[2] Institute for Morphological Endocrinology and Histochemistry, HHU Düsseldorf, D-40225 Düsseldorf, Germany
E-mail: Klaas@hirn.uni-duesseldorf.de, RK@hirn.uni-duesseldorf.de

Abstract

We introduce a formal method and its mathematical description for the translation of cortical maps. This method is based on three new classifications for cortical areas and projections between them and can easily be converted into algorithms for connectivity databases. As this approach works on logical, non-metric relations between parcellation schemes, in a minority of cases the translation leads to ambiguous results. However, we present several methods by which possible ambiguity can be minimized. Generally, this approach provides a framework for the translation of connectivity data in an objective and reproducible way and thus helps to overcome methodological difficulties of previous neuroanatomical databases.

Keywords: Parcellation scheme, brain map, cerebral cortex, cortical area, connectivity, analysis, database

1 The problem of incongruent cortical maps

Throughout this century, a variety of criteria have been proposed for the parcellation of the cerebral cortex into structurally and functionally homogenous units - the so-called "cortical areas." In the past a single structural criterion, like the laminar arrangement and the types of neurons (cytoarchitecture) or the orientation and density of myelinated fibres (myeloarchitecture), was usually applied to gain a parcellation scheme of the cortex. Nowadays, however, it is commonly agreed that only a combination of criteria (e.g. cyto- and myeloarchitecure, connectivity patterns, types of thalamic afferents, receptor and enzyme distribution, electrophysiological properties) can yield valid cortical maps. However, there is less agreement concerning the combinations of criteria that are most suitable and the methods that deliver the most reliable data in accordance to the chosen criteria (see Van Essen 1985 and Gahr 1997 for reviews). Thus, even using the same criterion, the growing arsenal of neuroanatomical and neurophysiological methods has led to a variety of different cortical maps for each species in this century. Furthermore, as some of these criteria (especially cytoarchitecture and myeloarchitecture) also strongly depend on the observer and the experimental conditions, an unanimous parcellation is seldom achieved even if criteria and methods agree.

The problem of comparing and translating different cortical parcellation schemes has hardly been tackled seriously so far (see Zilles 1990 for a review). One specific and increasingly more important aspect of the parcellation problem is the usage and analysis of connectivity data between cortical areas that rely on different parcellation schemes of the cerebral cortex. Throughout the last years, the notion that an understanding of the brain's principles of function is impossible without using mathematical tools of analysis and simulation applied to its highly complex global wiring pattern, has been endorsed by an increasing number of authors. Although some of these approaches have already led to new insights into the brain's possible overall organization (e.g. Felleman & Van Essen 1991, Young 1992, Scannell et al. 1995, McIntosh et al.1994, 1996, Hilgetag et al. 1996), the methodological question of how to deal with the different parcellation schemes on which the connectivity data is based has not been dealt with so far.

2 Description of the method

2.1 Three classifications as the basis of an algebra

Let us look at a practical example: Brodmann's area 7 (Brodmann 1905, 1909) covers large portions of both lateral and medial aspects of the parietal lobe. Other parcellation schemes, however, distinguish several subareas within this area. Von Bonin & Bailey (1947), for example, subdivide area 7 into areas PF, PG and PEp (see Figs. 1A and 1B).

Now imagine yourself dealing with connectivity data that refers to the parcellation scheme of von Bonin & Bailey. Assuming we have projections to another specific area originating from areas PF and PG, but not from PEp - how could this situation be described appropriately in Brodmann's map?

Much more difficult still is the situation when a brain map not only differs from another in subdividing or uniting areas of the latter, but presents completely different borders, so that areas in the two maps only partially overlap.

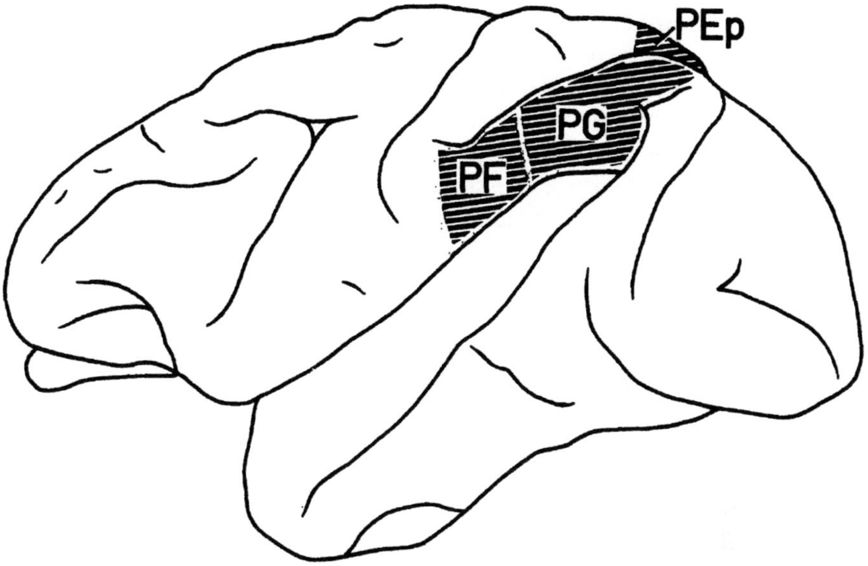

Fig. 1A: Parcellation of the cerebral cortex of the macaque, lateral aspect. Brodmann's area 7 (hatched) is equivalent to the combinations of areas PF, PG and PEp in von Bonin & Bailey's map

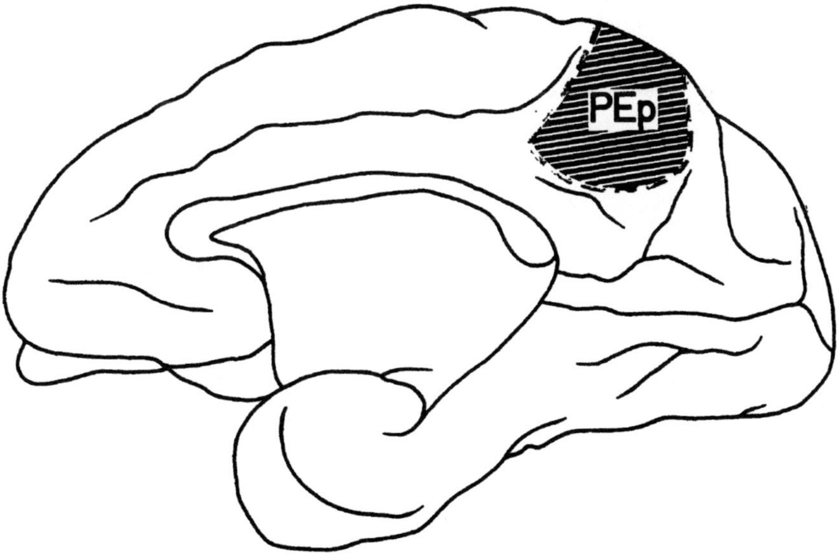

Fig. 1B: Parcellation of the cerebral cortex of the macaque, medial aspect. Brodmann's area 7 (hatched) is equivalent to the combinations of areas PF, PG and PEp in von Bonin & Bailey's map

To cope with difficulties like these, we have introduced three simple logical classifications which improve the translation of connectivity data between different brain maps considerably. Before they are described, a conceptual difficulty concerning the comparison of different brain maps has to be resolved first.

As not only brain maps differ, but are also presented on the basis of individual brains with a certain variance of morphology (that is, differences in overall shape, cortical folding and variable relations of areas to landmarks like sulci and gyri), it is difficult to compare cortical parcellation schemes without referring to a common standard. Throughout the following approach we will, therefore, refer to a "standard brain" which is thought to represent a normalized projection sheet for the different existing cortical maps. The reader may envisage this as a process in which an individual brain, on whose surface a particular parcellation scheme is displayed, is warped by non-linear transformations until it matches the standard brain perfectly.

2.1.1 Relation Codes (RC)

The first classification concerns the possible relations between two areas A and B of two different cortical maps which are projected simultaneously onto a "standard brain". We distinguish 4 possible relations that express a relative (logical), not an absolute (metric) degree of overlap:

- **Id** ("Identical area"): Areas A and B are identical.
- **S** ("Subarea"): A is a subarea of B, that is, it is completely contained in B.
- **L** ("Larger area"): The reverse of case S, that is, A completely contains B.
- **O** ("Overlapping area"): The two areas overlap; each of them covers some part of the standard cortex that is not contained in the other area.

In the following, these codes will be addressed as "relation codes" (RC).

2.1.2 Extension Codes (EC)

The second classification refers to the question whether the origin or termination of a particular projection spreads throughout the whole extent of an area or is restricted to some part of it:

- C ("Complete"): The whole area is the origin or the recipient of fibres of the projection.
- P ("Partial"): Only subparts of the area are the origin or the recipient of fibres of the projection.
- N ("None"): No part of the area is the origin or the recipient of fibres of the projection.

These codes will be addressed as "extension codes" (EC).

Every projection consists of a "sending" (or "source") and a "receiving" (or "target") area. Depending on whether an extension code refers to a sending or receiving area, it will be addressed as "sending code" (sC) or "receiving code" (rC), respectively.

2.1.3 Projection Codes (PrC)

Furthermore, by combining the sending and receiving code of a projection, we obtain a third classification which describes the properties of a particular projection as a whole. This two-letter-code will be called the "projection code" (PrC) and comprises 5 possible values:

- NN: No part of the source area projects to any part of the target area.
- PP: Subparts of the source area project to subparts of the target area.
- PC: Subparts of the source area projects to the whole extent of the target area.
- CP: The source area projects in its whole extent to subparts of the target area.
- CC: Fibres of the projection originate throughout the whole extent of the source area and terminate throughout the whole extent of the target area.

2.2 A first informal description of an algebra

On the basis of these three classifications it is now possible to set up an algebra for the translation of cortical connectivity data, that is, a set of rules which are applied when operating upon a combination of relation codes and projection codes. More specifically, these rules describe what relation and extension codes you obtain if one or two areas in the source map with particular relation codes and particular extension codes are translated to the target map. It is important to note that these rules are applicable to two areas at most in one step. Thus, if you have to consider more than two areas (like our example, in which area 7 was subdivided into PF, PG, PEp), you have to apply the appropriate rules to two areas with each step. After the first step, one of these areas is equivalent to the unification of the ones you have already dealt with. Therefore, in such a case of multi-step translation, apart from a resulting extension code in each step, you will also need a resulting relation code for the next step of translation. That is what these rules deliver.

Still rather informally at the moment, these rules are summarized by Table 1. Their exact mathematical properties will be explained later on.

Generally, the rules in this table describe the outcome of a single translation step and provide input, if necessary, to the next step. For example, rule 9 tells us that if we deal with two areas in our source map that are both subareas of another area in our target map (RCs S / S) and have extension codes of C and N (ECs C / N), then the output for the next step of translation is RC=S and EC=P. Notice that there are some rules which require only one RC/EC-pair as input and deliver just an extension code, but not a relation code, as output. These rules are linked to cases

Table 1: Informal summary of translation rules

Rule Nr	RC of 1st/2nd area	EC of 1st/2nd area	Resulting RC	Resulting EC
1	Id	C / -		C
2	Id	P / -		P
3	Id	N / -		N
4	L	C / -		C
5	L	P / -		N, P, C
6	L	N / -		N
7	S / S	C / C	S	C
8	S / S	C / P	S	P
9	S / S	C / N	S	P
10	S / S	P / P	S	P
11	S / S	P / N	S	P
12	S / S	N / N	S	N
13	S / O	C / C	O	C
14	S / O	C / P	O	P, C
15	S / O	C / N	O	P
16	S / O	P / C	O	P
17	S / O	P / P	O	P
18	S / O	P / N	O	P
19	S / O	N / C	O	P
20	S / O	N / P	O	N, P
21	S / O	N / N	O	N
22	O / O	C / C	O	C
23	O / O	C / P	O	P, C
24	O / O	C / N	O	P
25	O / O	P / P	O	N, P, C
26	O / O	P / N	O	N, P
27	O / O	N / N	O	N

in which areas of the source map possess a relation code of Id or L and can thus be translated in a single step.

Also, the rightmost column of the table shows that in some cases there is no unequivocal result. This happens when one of the areas from the source map is an overlapping or a larger area (RC=O / L) and, at the same time, one of the extension codes denotes a partial projection (EC=P). What is the reason for this? As there are no data available yet of the exact spatial relationship of two areas in two different maps (that is, their topographical relation if they were projected simultaneously onto a standard cortex) our method of translation can only operate on a logical, non-metric relationship between two such areas. If we now deal with a partial extension of fibres within one area of the source map and this area overlaps only partially with another area from the target map, we cannot say with certainty whether the subpart of the first area containing the fibres of the projection, lies within that part of our standard cortex that is covered by both areas. Depending on the position of the subpart containing the fibres and its extension, it might be situated completely outside the common overlap of the two areas, make up a part of it or even include it completely. The implications and possible solutions to this problem as well as the characteristics of the different kinds of rules and their formal definition will be given in detail below.

Before we turn to the formal description, however, let us look at some more practical examples which will help to understand how to read this table:

To start with an almost trivial case: If we consider Brodmann's area 17 and von Bonin & Bailey's area OC, we deal with a case of two areas in different maps that only bear different names but have identical borders. That is, their relation code is RC=Id. As we would expect, the first three rules of this table tell us that, whatever the extension code of one of these areas, it will stay the same in the process of translation.

Less obvious the case if we return to our initial example of Brodmann's area 7. Envisage three projections from PF, PG and PEp (as described by the map of von Bonin & Bailey) to OC with the projection codes CP, PP and NN (that is, the projection PEp→OC has been found to be absent). We will now only take a look at the translation of the sending codes and ignore the receiving codes at the moment. Knowing that PF, PG and PEp altogether correspond to area 7 (thus for each of them their relation code is RC=S), how would you describe the transfer of the connectivity data of PF, PG, PEp to area 7? Applying rules 8 and 9 of the algebra recursively, we first apply rule 8 to (S,C) and (S,P) (the data referring to PF→OC and PG→OC, respectively) and obtain (S,P). In other words, this means that the unification of PF and PG altogether has a relation code of RC=S and an extension code of EC=P. This temporary result is, together with (S,N) (referring to PEp→OC), one of the input values for rule 11 which delivers EC=P as the resulting sending code.

Now imagine the reverse case: You face a projection from area 7 to area 17 in the map of Brodmann whose projection code is CP, and you are supposed to translate this projection to the map of Bonin & Bailey. As the relation code of area 7 is RC=L (being the area that comprises PF, PG and PEp), we have to use rule 4 of the table which delivers a sending code of EC=C for each of our three areas PF, PG and PEp. This result is intuitively understood, since the fibres of the projection originate throughout the whole extent of area 7 and thus also from the whole extent of areas PF, PG and PEp. As this part of the translation is accomplished in just one step, this rule does not need to deliver a relation code.

2.3 A precise formulation of the new approach

Before we can turn towards a precise formulation of the problem of translating cortical maps and offer a possible solution, a conceptual prerequisite is necessary in order to make mathematical conventions applicable to cortical areas:

Within the following approach, we envisage a standard brain whose cortical surface is divided into a finite number of small patches each of which may be called an "elementary microarea" (EMA). These EMAs are thought to be small enough to lie within all the areas that would result from simultaneous projection of all existing cortical maps onto the standard cortex. Each cortical area of any cortical map is thus assumed to be equivalent to a specific set of a finite number of EMAs of the standard cortex (accordingly, each brain map is thus a set of a finite number of sets of a finite number of EMAs.) In this way, we have arrived at some formal units which are compatible with algebraic operations. If such an operation, like union or intersection, is applied to cortical areas defined this way, then this operation refers to the EMAs. For example, this convention allows us to describe two areas A and B of two different maps which share some part of the standard cortex, by the statement $A \cap B \neq \emptyset$. Similarly, the situation of an area A in one cortical map being equivalent to two smaller "subareas" B_1 and B_2 in another map can be expressed by $A = B_1 \cup B_2$.

The problem of translating connectivity data referring to different parcellation schemes can now be formulated generally:

Given two cortical maps A and B that consist of a and b areas, respectively:

(1) $$A = \{A_1, A_2, \ldots, A_a\}$$
$$B = \{B_1, B_2, \ldots, B_b\}.$$

In the following, we will denote projections between areas within these maps by one of the following ways:

In a simple form, a projection between two areas A_i and A_j can be described as $A_i \rightarrow A_j$ or as a simple tuple containing the two areas: (A_i, A_j). In a more complex form, we add information about the sending and the receiving code of the projection: $A_i \ {}^{sC}\!\!\rightarrow^{rC} A_j$ or written as a nested tuple: $(A_i, A_j, (sC, rC))$.

Now we envisage a specific connection from area A_{i0} to area A_{j0} that has a sending code sC and a receiving code rC:

(2) $$A_{i0} \ {}^{sC}\!\!\rightarrow^{rC} A_{j0} \qquad (1 \leq i_0 \leq a, \ 1 \leq j_0 \leq b)$$

or written as a tupel:

$$(A_{i0}, A_{j0}, (sC, rC))$$

To clarify the use of the nested index: In the following, we will use $i_0, i_1 \ldots i_{a-1}$ and $j_0, j_1 \ldots j_{a-1}$ as a set of indices with values from $\{1 \ldots a\}$ but with a different order, to denote the indices of areas from map A:

(3) $$\{i_0, i_1 \ldots i_{a-1}\} = \{1 \ldots a\}$$
$$\{j_0, j_1 \ldots j_{a-1}\} = \{1 \ldots a\}$$
$$\forall\, x, y \in \{0 \ldots a-1\},\, x \neq y:\ i_x \neq j_x \wedge i_x \neq i_y$$

For example, A_{i_2} thus does not necessarily denote area A_2, but definitely denotes a different area from map A than A_{j_2}. The reason for the choice of such a nested index will become obvious later on (see equations (6) and (7) especially).

The question now is: What is the equivalence of this particular projection $A_{i_0} \rightarrow A_{j_0}$ in the nomenclature of map B? In the following passages, we will describe a general method of how this problem can be solved:

First, we have to determine all areas within map B which share some parts of the standard cortex with A_{i_0} and A_{j_0}, respectively:

(4) $$saB = \{B_k \in B \mid 1 \leq k \leq b,\ B_k \cap A_{i_0} \neq \varnothing\}$$
$$taB = \{B_k \in B \mid 1 \leq k \leq b,\ B_k \cap A_{j_0} \neq \varnothing\}$$

In this way, saB is the set of those areas in map B that overlap in some way with A_{i_0} in the standard cortex. These areas will consequently be the source areas of those projections in map B that will result from the translation of $A_{i_0}\ {}^{sC}\!\!\rightarrow^{rC}\ A_{j_0}$ in map A. This is expressed by the abbreviated name of the set: saB = “source areas in map B“. Similarly, taB is the set of those areas in map B that overlap with A_{j_0} in map A (taB = “target areas in map B“) and thus contains the target areas of the resulting projections in map B.

The Cartesian product saB × taB then corresponds with the set of all connections within map B which will result from the translation of $A_{i_0}\ {}^{sC}\!\!\rightarrow^{rC}\ A_{j_0}$. For the sake of simplicity, this set will be adressed as ARC (“all resulting connections“) and is assumed to have $c \in N$ elements:

(5) $$\begin{aligned} ARC &= saB \times taB \\ &= \{(B_p, B_q) \mid 1 \leq p, q \leq b,\ B_p \cap A_{i_0} \neq \varnothing \wedge B_q \cap A_{j_0} \neq \varnothing\} \\ |ARC| &= c \qquad (c \leq b*(b-1)) \end{aligned}$$

Fig. 2 shows an example in which saB consists of areas B_1 and B_2:

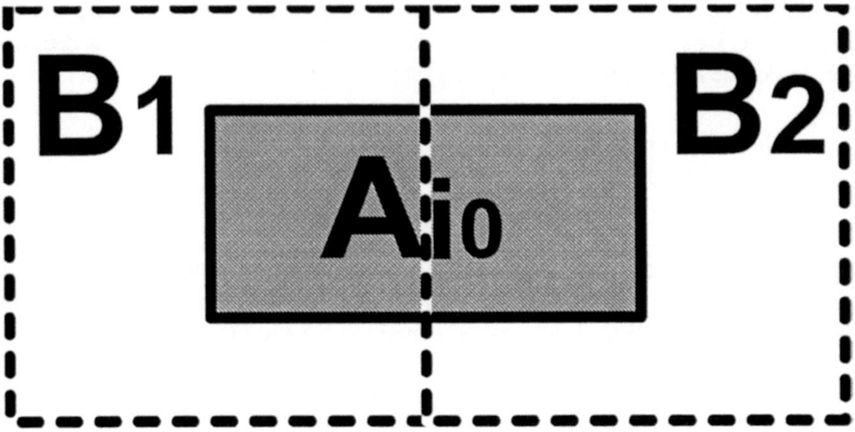

Fig. 2

For each resulting projection, that is for each element (B_p,B_q) of ARC, one now has to compute a projection code. That is feasible when not only the projection code of $A_{i0} \rightarrow A_{j0}$ is considered, but also the codes of all projections between those areas in A that, in addition to A_{i0} and A_{j0}, overlap with B_p or B_q, respectively. For each resulting projection (B_p,B_q), these areas can be described as

(6) $$saA = \{ A_{ik} \in A \mid 0 \leq k \leq a-1,\ A_{ik} \cap B_p \neq \varnothing \}$$
$$taA = \{ A_{jk} \in A \mid 0 \leq k \leq a-1,\ A_{jk} \cap B_q \neq \varnothing \}$$

(Similarly to saB and taB, saA means "source areas in map A" and taA means "target areas in map A".)

Then the following two statements apply to these sets:

(7a) $$\exists\, n \in N,\ 0 \leq n \leq a:$$
$$saA = \bigcup_{x=0}^{n-1} A_{ix} \supseteq B_p \wedge saA / \{A_{iy}\} \subset B_p \quad (0 \leq y \leq n-1,\ 1 \leq p \leq c)$$

(7b) $$\exists\, m \in N,\ 0 \leq m \leq a:$$
$$taA = \bigcup_{x=0}^{m-1} A_{jx} \supseteq B_q \wedge taA / \{A_{jy}\} \subset B_q \quad (0 \leq y \leq m-1,\ 1 \leq q \leq c)$$

Explanation in words for saA: It consists of n areas A_{i0} to A_{in-1} in map A which, as a unified piece of standard cortex, equal or include area B_p in map B. If any of these areas are left out, the resulting unified piece of cortex region lacks some part of the standard cortex that is included in B_p. The corresponding applies to taA.

Now the purpose of the nested indices also becomes obvious: For each resulting projection of the translation (that is, for each element of ARC), the values of the index sets $\{i_0, i_1 \ldots i_{a-1}\}$ and $\{j_0, j_1 \ldots j_{a-1}\}$ are set according to the areas of which saA and taA consist. Since the areas in these subsets of A are not necessarily in consecutive order, a simple index would not be adequate.

The Cartesian product saA × taA then corresponds with the set of all n*m projections within map A whose source areas overlap with B_p and whose target areas overlap with B_q:

(8) saA × taA =

$$\{(A_{i_x}, A_{j_y}) | A_{i_x} \in saA, A_{j_y} \in taA, B_p \cap A_{i_x} \neq \emptyset \wedge B_q \cap A_{j_y} \neq \emptyset\}$$

Note an important difference between saA × taA and saB × taB (=ARC): Whereas the latter has to be determined just once for the translation of a specific projection from map A, the former has to be determined anew for every projection in map B which results from this translation (and thus is an element of ARC). That means, that also the index sets $\{i_0, i_1 \ldots i_{a-1}\}$ and $\{j_0, j_1 \ldots j_{a-1}\}$ whose elements are determined according to the areas contained by saA and taA, respectively, are composed anew for each element of ARC.

Figure 3 gives an example of this important difference: Here saA is determined for a projection of ARC in which area B_2 is the source area (indicated by the shading). In this example, saA would comprise areas A_{i_0}, A_{i_2}, A_{i_3}. However, if saA was determined for a projection of ARC in which area B_1 was the source area, saA would comprise all four areas A_{i_0} - A_{i_3}.

Before we can actually compute the resulting projection codes belonging to the projections included in ARC from the projection codes in saA × taA, we will have to define the algebra for our translation method more accurately.

Generally, the definition of an algebra requires the specification of some sets and some functions that operate on members of these sets. In our case, these sets are the above defined relation codes (RC), extension codes (EC) and projection codes (PrC):

(9) RC = {Id, S, L, O}

EC = {N, P, C}

PrC = {NN, PP, PC, CP, CC}

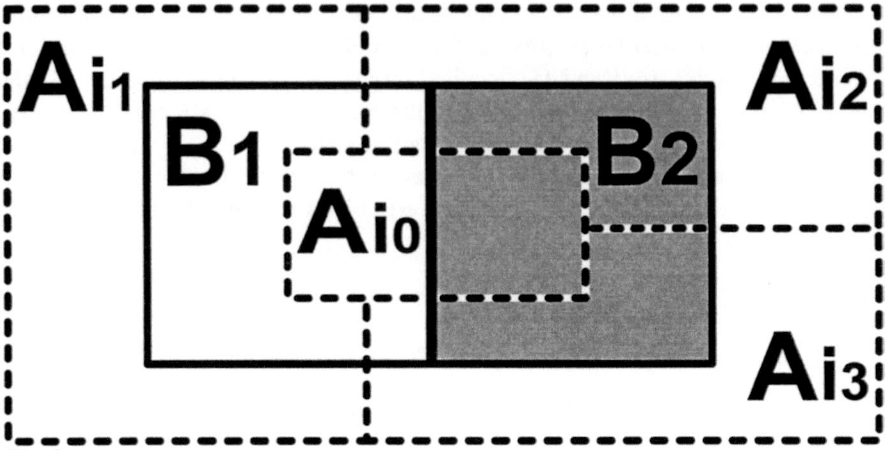

Fig. 3

Furthermore, we define three functions $\oplus_I$, $\oplus_T$ and Φ as (mathematical) projections between these sets:

(10)	$\oplus_I$: $(RC \times EC)^2 \rightarrow (RC \times EC)$	("intermediate" operation)
	$\oplus_T$: $(RC \times EC)^2 \rightarrow EC$	("terminal" operation)
	Φ: $(RC \times EC) \rightarrow EC$	("one-step" operation)

These three functions have the following characteristics:

The "intermediate" operation $\oplus_I$, which is used in multi-step translations for each but the final step, projects two pairs of an RC and an EC onto one such pair which then serves as one of the input values for the next step. The "terminal" operation $\oplus_T$ is used for the final step of a multi-step translation and delivers one EC as output. The "one-step" operation Φ needs just one such pair as input and, like $\oplus_T$, delivers one EC.

The exact nature of these three functions is shown by the following tables. As an analogy, it might be helpful to compare them to the lookup tables as delivered by Boolean algebra. The numbers assigned to the rules refer to the identical rules in Table 1. Some examples how to read these tables:

Table 2, rule 9: $\oplus_I$ ((S,C),(S,N)) = (S, P)
Table 3, rule 16: $\oplus_T$ ((S,P),(O,C)) = P
Table 4, rule 4: Φ (L,C) = C

In the following, we will use $(RC_1, EC_1) \oplus_I (RC_2, EC_2) = (RC_{res}, EC_{res})$ and $(RC_1, EC_1) \oplus_T (RC_2, EC_2) = EC_{res}$ as equivalent notations for $\oplus_I ((RC_1, EC_1), (RC_2, EC_2)) = (RC_{res}, EC_{res})$ and $\oplus_T ((RC_1, EC_1), (RC_2, EC_2)) = EC_{res}$, respectively.

Table 2: Function $\oplus_I$

Rule Nr	RC of 1st/2nd area	EC of 1st/2nd area	Resulting RC	Resulting EC
7	S / S	C / C	S	C
8	S / S	C / P	S	P
9	S / S	C / N	S	P
10	S / S	P / P	S	P
11	S / S	P / N	S	P
12	S / S	N / N	S	N
13	S / O	C / C	O	C
14	S / O	C / P	O	P, C
15	S / O	C / N	O	P
16	S / O	P / C	O	P
17	S / O	P / P	O	P
18	S / O	P / N	O	P
19	S / O	N / C	O	P
20	S / O	N / P	O	N, P
21	S / O	N / N	O	N
22	O / O	C / C	O	C
23	O / O	C / P	O	P, C
24	O / O	C / N	O	P
25	O / O	P / P	O	N, P, C
26	O / O	P / N	O	N, P
27	O / O	N / N	O	N

Now let us return to the process of the translation itself. Remember that having determined saA × taA, we face a number of n*m sending and receiving codes associated with the projections contained in saA × taA. Which of these extension codes should now be considered for computing the resulting projection code of (B_p,B_q)? Remember that saA × taA has to be determined anew for every single projection (B_p,B_q) of ARC and that all areas of saA and taA as a unified piece of cortex are equivalent to or include B_p and B_q, respectively. Any projection contained by saA × taA which has a sending or receiving code of EC=P is thus very likely to affect also B_p or B_q, respectively. If the extension code of such a projection is EC=C, there even is absolute certainty about this. Therefore, it seems reasonable that for each area contained in saA and taA, the maximum (that is, most extensive) sending and receiving code, respectively, of all projections which originate or terminate in this area and are an element of saA × taA should be chosen to represent this area:

$$(11)\quad \forall A_{ik} \in saA: sC_{ik} = \max\{s \mid (A_{ik}, A_{jx}, (s,r)) \in saA \times taA\} \quad (0 \le x \le m-1)$$
$$\forall A_{jk} \in taA: rC_{jk} = \max\{r \mid (A_{ix}, A_{jk}, (s,r)) \in saA \times taA\} \quad (0 \le x \le n-1)$$

Table 3: Function $\oplus_T$

Rule Nr	RC of 1st/2nd area	EC of 1st/2nd area	Resulting EC
7	S / S	C / C	C
8	S / S	C / P	P
9	S / S	C / N	P
10	S / S	P / P	P
11	S / S	P / N	P
12	S / S	N / N	N
13	S / O	C / C	C
14	S / O	C / P	P, C
15	S / O	C / N	P
16	S / O	P / C	P
17	S / O	P / P	P
18	S / O	P / N	P
19	S / O	N / C	P
20	S / O	N / P	N, P
21	S / O	N / N	N
22	O / O	C / C	C
23	O / O	C / P	P, C
24	O / O	C / N	P
25	O / O	P / P	N, P, C
26	O / O	P / N	N, P
27	O / O	N / N	N

Table 4: Function Φ

Rule Nr	RC	EC	Resulting EC
1	Id	C	C
2	Id	P	P
3	Id	N	N
4	L	C	C
5	L	P	N, P, C
6	L	N	N

sC_{ik} and rC_{jk} thus denote the maximum sending and receiving code for areas A_{ik} and A_{jk} as elements of saA and taA, respectively. In this way, we arrive at n sending codes and m receiving codes that will serve as input values for the operations of the algebra.

Further, let $sRC_{i0}, sRC_{i1} \ldots sRC_{in-1}$ denote the relation codes of the n areas contained in saA and $rRC_{i0}, rRC_{i1} \ldots rRC_{im-1}$ denote the relation codes for the m areas contained in taA.

Each of the areas contained by saA and taA is now associated with a maximum sending or receiving code, respectively, as well as with a relation code which indicates the relation between this area and B_p or B_q, respectively.

Now we can recursively apply the operations delivered by the algebra to the areas contained in saA and taA:

If we deal with a relation code of RC=O or RC=S we possibly need more than one step to accomplish the translation. In this case, both the resulting connection code and the resulting relation code of one translation step represent input values for the next step. The whole sequence by which the resulting sending and receiving codes are computed, can be expressed as:

(12) $sC_{res} =$

$$(...(((sRC_{i0}, sC_{i0}) \oplus_I (sRC_{i1}, sC_{i1})) \oplus_I (sRC_{i2}, sC_{i2}))...) \oplus_T (sRC_{in-1}, sC_{in-1})$$

$rC_{res} =$

$$(...(((rRC_{j0}, rC_{j0}) \oplus_I (rRC_{j1}, rC_{j1})) \oplus_I (rRC_{j2}, rC_{j2}))...) \oplus_T (rRC_{jm-1}, rC_{jm-1})$$

If the relation code is RC=Id or RC=L, however, we have only one pair of input values and a single translation step:

(13) $sC_{res} = \Phi\,(sRC, sC)$

$rC_{res} = \Phi\,(rRC, rC)$

In words, this means that, whenever we deal with an area whose relation code is RC=S or RC=O, we perform the "intermediate" operation $\oplus_I$ in a step-by-step process for all the areas contained in saA and taA, respectively. The result of each of these steps is part of the input for the next step, and so on, until we deal with the last area contained in saA or taA. Then the "terminal" operation $\oplus_T$ is perfomed, which delivers the final sending or receiving code. This also illustrates why we need two different operations in this case: The output of $\oplus_I$ has to include information on the overall relation that the combined areas of the source map, which have been considered in the translation process so far, have to the overlapping or including area in the target map. Otherwise, the translation could not be continued. $\oplus_T$, on the other hand, is responsible for the final step, that is, the computation of the resulting sending or receiving code.

If, however, the relation code is RC=Id or RC=L, we are faced with a considerably easier case: The translation can be performed in a single step, using operation Φ of the algebra.

2.4 A simple example

Taking up again the simple example that we already dealt with, we illustrate how the whole translation process can be described formally with the definitions of the classifications and the operations of the algebra:

Envisage a projection from PEp to OC (as described by the map of von Bonin & Bailey) which has been reported to be absent, that is, its projection code is PrC_1=NN. How can we translate this observation into the nomenclature of Brodmann?

Common sense might offer the quick suggestion that all one has to do is to describe the projection from area 7 to area 17 as absent, that is $7^{N}\rightarrow^{N}17$. However, this is not the case: In a first step, we determine all areas in the Brodmann map that overlap with these areas. Since PEp is a subarea of area 7 (and therefore sRC_1=S) and OC is equivalent to area 17 (and thus rRC_1=Id), we have

saB = {7}
taB = {17}.

Thus, the translation of PEp→OC will yield the single projection 7→17 in the map of Brodmann:

ARC = saB × taB = {(7, 17)}

Now, we must look for other areas in the map of von Bonin & Bailey which have to be considered for this translation. We know that, in addition to PEp, PF and PG also are subareas of area 7:

sRC2,3 = S
saA = {PEp, PF, PG}
taA = {OC}

Consequently, the projection codes of the following projections have to be considered:

saA × taA = { (PEp,OC), (PF,OC), (PG,OC) }

In our example, the two projections PF→OC and PG→OC have the projection codes CP and PP. The next step is to determine from these projection codes the maximum sending and receiving codes for each area contained in saA and taA:

sC1 = N, sC2 = C, sC3 = P, rC1 = P.

We can now compute the resulting projection code for 7→17. Dealing first with the resulting sending code of the projection, we have the following equation:

$$\begin{aligned} sC_{res} &= ((RC_1, sC_1) \oplus_I (RC_2, sC_2)) \oplus_T (RC_3, sC_3) \\ &= ((S,N) \oplus_I (S,C)) \oplus_T (S,P) \\ &= (S,P) \oplus_T (S,P) \\ &= P \end{aligned}$$

As for the resulting receiving code, we obtain

$$\begin{aligned} rC_{res} &= \Phi\,(RC_1, rC_1) \\ &= \Phi\,(Id, P) \\ &= P \end{aligned}$$

Thus, the resulting projection code would be PrC_{res}=PP and the completely translated projection in Brodmann's map would be described as $7 \overset{P}{\rightarrow}{}^{P} 17$ or (7, 17, (P,P)).

Although this is a very simple example, it is easy to imagine that there are far more complicated situations which cannot be solved without a solid formal basis for an algorithmic procedure that guarantees objective, reproducible results.

2.5 Possible solutions to remaining difficulties

Finally, we will demonstrate possible ways of how to cope with the remaining difficulties of this approach.

The most obvious remaining problem is probably that of ambiguous output of algebra operations in some cases (see chapter 2.4). This problem is not as serious as it might seem to be at a first glance. First, only 4 of the 27 cases described by Table 1 lead to ambiguous results in which a distinction between existing projections (that is, the extension code resulting from the translation is EC=P or EC=C) and absent projections (that is, the resulting extension code is EC=N) is not possible. Furthermore, 3 of these 4 cases occur only if areas of different maps overlap (that is, in association with a relation code of RC=O). Our experiences with relations between brain maps, however, show that this case is quite rare and that brain maps predominantly differ by subdividing existing areas. In a recent analysis (Hilgetag et al. 1997), for which this approach was used to translate connectivity data based on four different cortical maps, only 16% of the relationships between areas of different maps were of this type.

More important than statistical factors, however, are a variety of methods which can be applied to effectively minimize or completely eliminate ambiguity:

First, as we usually deal with a large number of different maps and for each of these maps information about the relation to not just one, but several other maps is available, we can represent the areas of the different maps as a complex graph, its nodes being the areas and its edges being the known relations. By employing powerful algorithms from graph theory (e.g. the so-called "algorithm of Dijkstra"), one can then decide for each translation which way along the nodes of the graph implies the smallest cost, that is, the least ambiguity. In this way, ambiguity of the whole translation process can be reduced to a minimum.

Second, the processing of "safe cases" (that is, relation codes RC=Id and RC=S) may deliver information that can subsequently be used to decide about ambiguous situations. For example, this already is the case if we have only one single area in the source map that is a subarea of an area in the target map. Take a look at Figure

4 and imagine that you deal with the translation of a projection from areas A_1-A_4 in the source map to area B in the target map. As you will notice immediately, areas A_1 and A_4 are overlapping with B, thus they could attribute ambiguity to the overall translation (see rules associated with RC=O of Tables 1-3). But if the extension code of one of the areas A_2-A_3 is other than "N", insecurity is reduced considerably: We then know for sure that the resulting extension code can by no means be EC_{res}=N. If one of the areas A_2-A_3 has an extension code of EC=P, we even know immediately, that the resulting extension codes must be EC_{res}=P, irrespective of the extension and relation codes of the other areas. Generally, a single subarea with a partial extension code is sufficient to eliminate ambiguity caused by any number of overlapping areas.

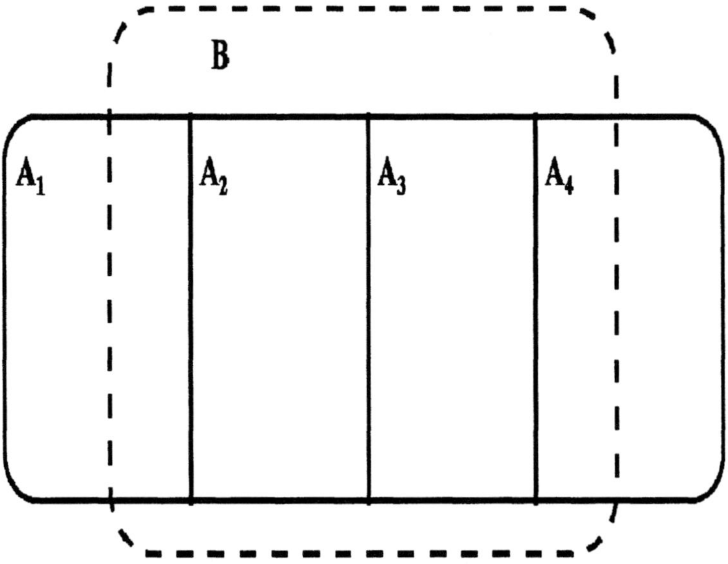

Fig. 4

A third and also very powerful method to effectively reduce ambiguity is to use other unambiguous reports for verification: Usually, a database of raw connectivity data will contain a rather big amount of redundancy. That means, that several reports using different nomenclatures present similar or identical information. We can thus look for such (partially or entirely) overlapping reports whose translation is unambiguous and use them do decide our ambiguous case. To use our example again: If we want to translate a projection $7^P \rightarrow ^P 17$ from Brodmann's map to an imaginary map X that subdivides area 7 into two parts (e.g. 7a and 7b) we would thus have to deal with a relation code of RC=L and a sending code of sC=P. Then, we would not be able to determine the resulting sending code for 7a and 7b unambiguously (see rule 5, Table 4). But, as we know that Brodmann's area 7 is made up of von Bonin & Bailey's areas PEp, PF, PG, we can look up the respective connections and relations in a database to get some help for our decision. If, for example, PEp is equivalent to 7a and PF and PG together are equivalent to 7b and we deal with the same projection codes as used in former instances of this example (that is, $PEp^{N} \rightarrow {}^{N}OC$, $PF^{C} \rightarrow {}^{P}OC$, $PG^{P} \rightarrow {}^{P}OC$), then we can use this information for an unambiguous translation from Brodmann's map to map X: We would then know with certainty that 7b, but not 7a is the origin of the projection (that is, the sending codes of 7a and 7b would be "N" and "P", respectively).

Generally, this means that within the framework of a connectivity database, we are always able to search for parcellation schemes which suggest subareas of the areas in question and use their respective connectivity data to come to an unambiguous decision.

On the whole, the combination of these methods ensure that possible ambiguity of the results of the translation process is effectively reduced to a minimum.

Another problem is much less obvious, but may nevertheless have been noticed by some readers: In those cases when the translation of a projection cannot be accomplished in one step, is the order in which the operations of the algebra are applied of any importance? Or, in mathematical terms: Are the algebra's operations commutative? The answer is neither a simple Yes nor No. The operations are commutative if all results of intermediate steps are unambiguous. If, however, an intermediate step should deliver a result whose ambiguity should not be overcome by the methods mentioned above, the operations of the algebra are not commutative. In this case, there are several possible ways in which the translation sequence might develop. This variety can be visualized by several (mathematical) trees with different structure and possibly different leaves (that is, results).

Although the operations are not commutative only in the case that there should remain some ambiguity in spite of our various methods to eliminate it, one nevertheless should carefully avoid the unlikely, but possible influence of the order in which operations are applied. It is therefore advisable to choose a fixed order of relation codes according to which the areas are chosen at each step of the translation. With such a fixed order, reproducibility of the translation is ensured. It seems advantageous to deal with safe cases (RC=Id and RC=S) first, since this would be compatible with one of the methods explained above for ambiguity elimination: Information gathered by the processing of subareas can be used to eliminate ambiguity produced by the more problematic overlapping areas (RC=O).

3 Discussion: Advantages and possible applications

It has to be emphasized again that there is an increasing agreement about the necessity of new ways of access to the huge amount of data that neuroanatomy has produced over the last decades. In contrast to disciplines like molecular biology in which the enormous progress is partly due to the fact that the unified knowledge about the genome is easily accessible by the help of powerful databases, brain research still lacks similar tools without which higher-order analysis of many aspects of the brain will not be possible. As one response to this situation, big efforts have been done during the last years and are still being done to build databases of connectivity data for several species (Felleman & Van Essen 1991, Scannell et al. 1995, Burns 1997). These databases allow access to the results of neuroanatomical tracer studies that have been accomplished for over twenty years and revealed an enormous number of connections between different parts of the brain (e.g. in the case of the rat (Burns 1997) more than 11.000 connection reports from more than 1.000 papers are stored and made accessible for all kinds of analytical tools).

The process of collating this data on brain connectivity is seriously complicated by the fact that the different authors use a wealth of different parcellation schemes. The existing databases are therefore founded on one compromise brain map chosen before the collation of the data was started. Thus, they do not contain raw data, but interpreted data. This constraint implicates several problems:

- Connectivity data of other than the chosen brain map had to be adjusted "by hand", a process whose results are neither objective (that is, independent of the individual observer) nor reliable (that is, constant in one observer over several times of data analysis).
- If errors or new insights concerning the interpretation of parcellation schemes are discovered, this leads to the need for corrections of the entered data themselves, possibly involving massive portions of them.
- If the choice of the chosen parcellation scheme turns out to be inadequate for particular purposes of analysis, there is no other way to change it, but to actually change all of the entered data accordingly.

The new approach presented above, however, avoids all of these problems and offers some attractive advantages:

- It allows us to enter the raw data of the papers in whatever parcellation scheme the author chose to present his results. Thus, the process of data collation becomes much more objective and reliable, since the author's original nomenclature can be used.
- The additional information (extension and relation codes) which it demands, is classified into few, clearly defined classes for which objective criteria can be formulated to a large degree. Besides, it is relatively easily obtainable (e.g. extension codes can be inferred from all figures with clear description of areal borders). If there is a lack of data for the relations between areas of two different maps, one can use methods from graph theory to find the shortest and least problematic way of translation via some intermediate maps. (A simple example: If we have no information about the relationship between maps A and B, but about the relationships between A and C on the

one hand, and B and C on the other hand, we can first translate A to C, then C to B.)

- Errors concerning the relationships discovered after the collation of data do no longer enforce changes of the raw connectivity data itself, but only lead to restricted changes to the table which describes the known relationships between maps.
- The user of the connectivity database is no longer forced to stick to one specific parcellation scheme. Instead, the user can choose among all parcellation schemes, whose relations to other maps are stored in the database, the one that is most suitable for the desired kind of analysis and even define new maps that represent combinations of other maps.

The optimal solution of the problem of diverging brain maps would be an atlas of a standard brain with three-dimensional coordinates which all data about individual brains would be referring to. Although some efforts are already done to construct such a tool (e.g. Roland & Zilles 1994), it will still take a long time until it will be ready to use. We believe that, until such a method is available, our approach will facilitate the collation of connectivity data for analysis considerably and will make neuroanatomical databases more versatile, flexible to use, and less prone to errors.

4 Acknowledgments

We thank F. T. Sommer, C.-C. Hilgetag, J.W. Scannell and K. Zilles for valuable discussions as well as for help with the manuscript. K.E.S. is supported by the Studienstiftung des Deutschen Volkes.

5 References

1. Bonin, G. von; Bailey, P.: The neocortex of Macaca mulatta. Urbana: University of Illinois Press 1947
2. Brodmann, K.: Beiträge zur histologischen Lokalisationd der Grosshirnrinde. Dritte Mitteilung: Die Rindenfelder der niederen Affen. J. Psychol. Neurol. 4:177-226 (1905)
3. Brodmann, K.: Vergleichende Lokalisationslehre der Grosshirnrinde in ihren Prinzipien dargestellt auf Grund des Zellenbaues. Barth: Leipzig 1909
4. Burns, G. A. P. C.: Neural connectivity of the rat: Theory, methods and applications, DPhil Thesis, University of Oxford 1997
5. Felleman, D.J.; Van Essen, D.C.: Distributed hierarchical processing in the primate cerebral cortex. Cereb. Cortex 1:1-47 (1991)

6. Gahr, M.: How should brain nuclei be subdivided? Consequences for developmental mechanisms and for correlations of area size, neuron numbers and functions of brain nuclei. Trends Neurosci. 20:58-62 (1997)
7. Hilgetag, C.-C.; O'Neill, M.A.; Young, M.P.: Indeterminate organization of the visual system. Science 271:776-777 (1996)
8. Hilgetag, C.-C.; Stephan, K.E.; Burns, G.A.P.C.; O'Neill, M.A.; Young, M.P., Zilles, K.; Kötter, R.: Systems of primate cortical areas defined by functional connectivity. Soc. Neurosci. Abstr. 23:514.13 (1997)
9. McIntosh, A.R.; Grady, C.L.; Ungerleider, L.G.; Haxby, J.V.; Rapoport, S.I.; Horwitz, B.: Network Analysis of cortical visual pathways mapped with PET. J. Neurosci. 14:655-666 (1994)
10. McIntosh, A.R.; Grady, C. L.; Haxby; J.V.; Ungerleider, L.G.; Horwitz, B.: Changes in limbic and prefrontal unctional interactions in a working memory task for faces. Cereb. Cortex 6:571-584 (1996)
11. Roland, P.E.; Zilles, K.: Brain atlases - a new research tool. Trends Neurosci. 17:458-467 (1994)
12. Scannell, J.W.; Blakemore, C.; Young, M.P.: Analysis of connectivity in the cat cerebral cortex. J. Neurosci. 6:3655-3668 (1995)
13. Van Essen, D.C.: Functional organization of primate visual cortex. In: Jones, E.G.; Peters, A (eds.).: Cerebral Cortex. Vol. III: Visual cortex. New York-London: Plenum Press 1985, pp. 259-329
14. Young, M.P.: Objective analysis of the topological organization of the primate cortical visual system. Nature 358:152-155 (1992)
15. Zilles, K.: Cortex. In: Paxinos, G. (ed.): The Human Nervous System. San Diego, Academic Press 1990, pp. 757-802

From Self-Organized Fluid Machines to Living Liquid Crystals

Hans Gruler: *Department of bioPhysics, Complex Fluids Group, University of Ulm, D 89069 Ulm, Germany and Centre d'Écologie Cellulaire, Hôpital de la Salpétrière, F 75651 Paris Cedex 13, France*

Abstract: The self-organization of amoeboid cells is discussed. The ordered state of amoeboid cells is produced out of a uniform state by an instability of the intracellular signal chain. The temporal pattern makes the the machine cycle while the spatial pattern produces the spatial spread machine: (i) an inactivated cellular state is predicted for low pumping; (ii) an isotropic mode for higher pumping (cell adhesion but no migration); (iii) a polar mode for even higher pumping (adhesion and directed migration); (iv) a bipolar activated cellular mode for even higher pumping (cell elongation and orientation); The machine characteristics of the different activated modes are discussed. An automatic controller (= controller with feedback) is responsible for the angle of migration and for angle of orientation. Migrating and interacting cells can form condensed states: A polar nematic liquid crystal is found for cells in the polar mode and an apolar nematic liquid crystal for cells in the bipolar mode.
Key-words: Chemo- and galvanotaxis, automatic controller, self-organization

Introduction

Cell movement and growth are important in wound healing, tumor invasion and metastasis, infection, inflammation, and embryogenesis including brain formation, etc. The behavior of the cells is determined by the stimuli coming from the environment of the cell and by the past history of the cell (heredity and memory). Directed cellular movement has been observed in a variety of motile cell types responding to various guiding fields in their environment for instance concentration gradient of chemoattractant molecules, electric field, chemical and physical modulations of the substrate, etc. The cellular response has been characterized as chemotaxis, necrotaxis, galvanotaxis, haptotaxis, contact guidance, etc. [1]. The involved chemistry is known to a large extent but the physical principles of how these cellular machines work

are less understood. The chief importance of this article will be the physical and technical aspects of cell migration, of cell growth, and of pattern formation of interacting cells. The "fruit fly" for our investigations are human granulocytes since they are easy to obtain and have great physiological importance [2]. Besides granulocytes other cell types were investigated like monocytes, lymphocytes, melanocytes, fibroblasts, osteoblasts, neural crest cells to name a few and also growing hyphen and nerve cones. However, the shown procedures are not restricted to these cell types.
Next, the self-organized structure of amoeboid cells will discussed.

Self-Organization of Amoeboid Cells

Many natural scientists are fascinated by the ability of amoeboid cells to form out of a uniform state self-organized machines. The physical concept for self-organization is the nonlinear dissipative dynamics as emphasized by Haken [3]. This concept is successfully applied to amoeboid cells [4, 5]. One important experimental fact for the modeling is that the machinery which creates the cell locomotion contains only a few elements like a part of the plasma membrane, unstructured cytoplasm (as seen by the light microscope), and the necessary biochemistry [6]. The spontaneous formation of temporal and spatial macroscopic structure is induced by the nonlinear dynamics in the cellular signal transduction chain [4, 5].
The involved cellular signal chain is as follows [7]: The first intracellular signal is created by the membrane-bound receptors activated by kinesis stimulating molecules of the extracellular space. The second intracellular signal (like inositol triphosphate, diacylglycerol, Ca^{++}, ...) is created by a chemical amplification process (G-protein, phospholipase C, ...). The cellular reaction units (adhesion proteins, actin, myosin, ...) act accordingly to the second intracellular signal. Diffusion processes in the signal chain are overwhelmed by a positive feedback which is basically the supply of fresh receptors induced by the second intracellular signal [4]. The formation of a spatial-temporal pattern of the second intracellular signal molecules is possible due to the feedback in the signal chain. The different self-organized fluid machines of amoeboid cells can be explained by this spontaneously formed spatial-temporal pattern [5] where the machine cycle is given by the temporal pattern [4] and the different cellular machines by the spatial pattern [5].
The action of a cell can be compared with that of a laser. For

a weak pumping, the action of the muscle proteins is weak and random and 'no' cellular action is expected (= inactivated cellular state). But for a strong pumping the action of the muscle proteins is strong and coherent even when the pumping is random, a coherent cellular action is expected (= activated cellular state). This self-organization concept is verified by experiments [4, 5]. The different self-organized cellular machines are the modes of the spatial pattern of the second intracellular messenger. The morphology of some amoeboid in different modes is shown in ref. [8]: (i) an amoeboid cell in the inactivated state is spherical with some membrane raffling; (ii) In the isotropic activated mode, the cell body is surrounded by a leading front containing oriented and coherently working muscle proteins but no direction is preferred. The cells in the isotropic mode have the ability to make adherence to a substrate. (iii) In the polar activated mode, the leading front is located at one side of the cell body. The cells in polar mode have the ability of directed and no-directed migration; (iii) In the bipolar activated mode, the leading front is located at two opposing sides of the cell body. The cells have the ability to elongate and orient their long axis.
Next, the machine properties of the amoeboid cells at different modes are discussed.

Cellular Automatic Controllers

There are different levels of understanding of biological phenomena. Most biologists would like to know all of the molecular steps in a given signal transduction pathway when describing a response mechanism. However, at the phenomenological level, a given response can be thought of as being controlled by an automatic controller, and the properties of this controller can be studied and modeled, providing some insight into the nature of the response mechanism without knowing all of the molecular steps involved [9]. Servo mechanisms of the technical world are control systems in which a hydraulic, pneumatic, electric or other type of controlling mechanism is actuated and controlled by a low energy signal. In biological systems, servo mechanisms are based on chemical reactions.
The hallmarks of a cellular automatic control are twofold; That is an element which measures the cellular output like the angle of migration and a means of feeding back this information into the input in such a way to minimize the deviation of the output from the desired level. A schematic representation of an auto-

matic controller is shown in [10] (Fig. 2b) where the controlled variable is the angle of migration, φ. In the detection unit the actual angle of migration is measured in the signal transformer characterized by $g_1(\varphi, E_1)$ and compared with the desired one, φ_0, in the summation point. The signal, M_1, which leaves the detection unit, should be a function of the applied extracellular guiding field, E_1, and of the deviation between the actual angle and the desired one. The signal of the detection unit enters the reaction unit characterized by k_1 which is composed of a chemical amplifier and a rotational motor.

The existence of a cellular automatic control mechanism can be shown by performing guiding signal-jump experiments since in the case of an automatic control mechanism the cell follows automatically the temporal variations of the guiding signal. For example, in the case of the polar cellular mode: First, the cells are allowed to migrate in an extracellular guiding signal during a time period ($t < 0$), and the directed migration can be quantified by a polar order parameter, $\langle\cos\varphi\rangle_0 \neq 0$. Then, at time $t = 0$ the direction of the extracellular guiding signal is reversed. The cells having an automatic controller react to the altered extracellular signal. Granulocytes for example make a U-turn towards the new guiding signal direction [11]. The polar order parameter reaches a new saturation value ($\langle\cos\varphi\rangle(t) \to -\langle\cos\varphi\rangle_0$). This behavior of the automatic controller can be predicted and is, thus, described by a deterministic signal ($g_1(\varphi, E_1)\, k_1$) in the automatic controller. But, besides the deterministic signal there exist a stochastic signal in the automatic controller, $\Gamma_1(t)$, since in case of no extracellular guiding signal, $E_1 = 0$, the cellular controller creates random changes in the migration direction ($d\varphi/dt \neq 0$). Thus, the cellular automatic controller of the polar mode of amoeboid cells must be characterized by a stochastic differential equation

$$\frac{d\varphi}{dt} = -g_1(\varphi, E_1) \cdot k_1 + \Gamma_1(t) \tag{1}$$

To proceed further one has to know some details of the stochastic processes in the automatic controller. For simplicity, $\Gamma_1(t)$ is approximated by a white noise of the strength q_1 ($\langle\Gamma_1\rangle = 0$, $\langle\Gamma_1(t)\,\Gamma_1(t')\rangle = q_1\,\delta(t-t')$). The proposed mathematical structure describing the cellular automata is actually verified by experiments [12].

Similar experiments can be performed with amoeboid cells in the bipolar mode. The orientation of the elongated cells is performed by an automatic controller which is described by a stochastic differential equation

$$\frac{d\Theta}{dt} = -g_2(\Theta, E_2) \cdot k_2 + \Gamma_2(t) \tag{2}$$

The orientation angle, Θ, of the bipolar cellular mode is measured in respect to the perpendicular direction of the extracellular guiding signal.
Due to the stochastic properties in the cellular signal transduction chain, the migration angle and the angle of orientation cannot be predicted, but their statistical properties can. The probability that the angle of migration falls between φ and $\varphi + \Delta\varphi$ is described by the angular distribution function $f(\varphi)$. The distribution functions $f_1(\varphi)$ and $f_2(\Theta)$ can be obtained from the stochastic differential equations [13] (Eqs. 1 and 2)

$$f_1(\varphi) = f_{10} \exp\left(\frac{2\,k_1}{q_1} \int g_1(\varphi, E_1)\, d\varphi\right) \tag{3}$$

$$f_1(\Theta) = f_{20} \exp\left(\frac{2\,k_2}{q_2} \int g_2(\Theta, E_2)\, d\Theta\right) \tag{4}$$

The angle distribution function, $f_1(\varphi)$, of migration cells, and the angle distribution function, $f_2(\Theta)$, of oriented cells can be determined from experiments and compared with the theoretical predictions.
Next, details of the cellular automatic controllers are determined.

Detection and Reaction Unit

Directed Migration: The angle of migration can be determined from trajectories. All the measured angles are collected in an angular distribution function, $f_1(\varphi)$, and compared with the predictions.
The mathematical structure for the cellular detection unit can be derived by using the symmetry of the system: $g_1(\varphi, E_1) = E_1^n \cdot \sin\varphi \pm \cdots$ with odd n. The angular dependence of the detection unit can be obtained by plotting the logarithm of the angular distribution function, $f_1(\varphi)$, versus $\cos\varphi$. A straight line is observed as predicted [9]. This holds for many cell types. The sensitivity of the cellular signal transformer, E_1^n, is obtained by investigating the signal-dependence of the angular distribution function. A broad distribution is obtained for small signal strengths and a narrow one for large signal strengths. One finds for many

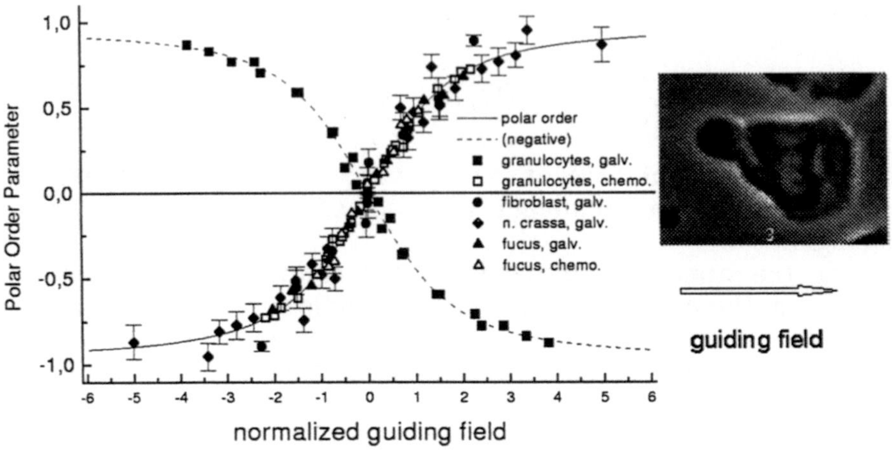

Figure 1: *The dose-response curve for directed migration and growth. On the right side a granulocyte in the polar mode (phase contrast) is shown.*

cell types that the coefficient, a_1, which characterize the width of the distribution is proportional to the strength of the guiding signal (n=1). One has in the case of an electric guiding signal, E, (galvanotaxis)

$$a_1 = \frac{2k_1}{q_1} \cdot E_1 = K_G\, E \tag{5}$$

where the cellular signal sensitivity is described by the galvanotaxis coefficient, K_G. Typical value of K_G^{-1} is 0.2 V/mm (granulocytes). One has in the case of a concentration gradient (chemotaxis)

$$a_1 = \frac{2k_1}{q_1} \cdot E_1 = K_{CT}^0 \frac{K_R}{(c + K_R)^2} \frac{dc}{dx} \tag{6}$$

with K_R for the binding of the chemoattractant to the membrane-bound receptor and the chemotaxis coefficient K_{CT}. A typical value for K_{CT}^0 is 9 mm for granulocytes.
The steady state dose-response curve for directed migration or growth is shown in Fig. 1. It is a dimensionless representation which characterizes the steady state response with respect to a polar guiding signal. The dimensionless polar order parameter which characterizes the cellular response is plotted versus the dimensionless polar guiding signal. Note, it is a fit-free representation if the galvanotaxis coefficient, K_G, and the chemotaxis coefficient, K_{CT}

are known. The solid line is a theoretical prediction (derived from Eq. 1). The dots represent actual measurements. The average of $\cos\varphi$ (polar order parameter) is calculated from the angular distribution function and the dimensionless guiding field is obtained by multiplying e.g. the applied electric field with the galvanotaxis coefficient.
We have demonstrated that the cellular machine responsible for the direction of migration, is an automatic controller. But the question of which type of controller is used by the cell is not yet answered. In the technical world a proportional controller is often used for simple problems. A proportional controller derives its name from its response since the output signal is just proportional to the input signal which is the difference between the desired level and the actual one. $\sin\varphi \approx \varphi$, thus, for small angles the cellular response is proportional to the deviation between the desired and actual angle. Amoeboid cells in the polar mode have a proportional controller for the directed migration.
We have demonstrated that the self-organized structure of amoeboid cells acts as an automatic controller. But, which essential signal is used is not yet answered?. The essential signal in the cellular detection unit is the dipole moment of the spatial distribution of the loaded membrane-bound receptors. It is the asymmetric loading of the membrane-bound receptors in case of chemotaxis and the asymmetric supply of membrane-bound receptors in the case of galvanotaxis. The polar mode of an amoeboid cell acts as an automatic controllers where the symmetry axis of the polar mode is altered in small steps by means of the dipole moment of the distribution of the loaded membrane-bound receptors.

Cell Orientation: The angle of orientation can be determined from photographs. All the measured angles are collected in an angular distribution function, $f_2(\Theta)$ and are compared with the predictions.
The mathematical structure for the cellular detection unit can be derived by using the symmetry of the system: $g_2(\Theta, E_2) = E_2^n \cdot \sin 2\Theta \pm \cdots$ with even n. The angular dependence of the detection unit can be obtained by plotting the logarithm of the angular distribution function, $f_2(\Theta)$, versus $\cos 2\Theta$. A straight line is observed for many cell types as predicted [14]. The sensitivity of the cellular signal transformer, E_2^n, is obtained by investigating the signal-dependence of the angular distribution function. A broad distribution is obtained for small signal strengths and a narrow one for large signal strengths. One finds for many cell types that the coefficient, a_2, which characterize the width of the distribu-

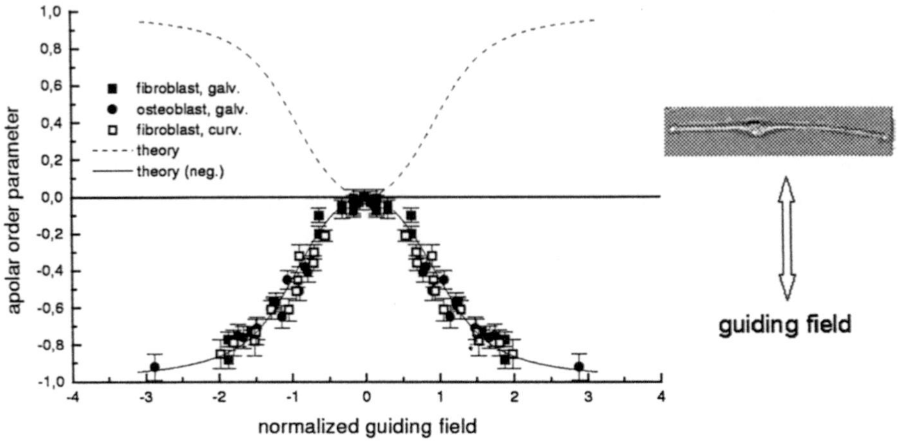

Figure 2: *Dose-response curve for cell orientation. On the right side a melanocyte in the bipolar mode (phase contrast) is shown.*

tion, is proportional to the square of the strength of the guiding signal (n=2) [14]. One has in the case of an electric guiding signal, E,

$$a_2 = \frac{2k_2}{q_2} \cdot E_2^2 = (K_2\, E)^2 \tag{7}$$

where the cellular signal sensitivity is described by the coefficient, K_2. Typical values of K_2^{-1} are 0.2 V/mm (fibroblasts). One has in the case of a bent surface (cylinder with curvature C)

$$a_2 = \frac{2k_2}{q_2} \cdot E_2^2 = (K_C\, C)^2 \tag{8}$$

A typical value for K_C is 85 μm for fibroblasts and osteoblasts. The steady state dose-response curve for cell orientation is shown in Fig. 2. It is a dimensionless representation which characterizes the steady state response with respect to an apolar guiding signal. The dimensionless apolar order parameter which characterizes the cellular response, is plotted versus the dimensionless guiding signal. Note, it is a fit-free representation if the coefficients, K_2, K_C, etc are known. The solid line is a theoretical prediction (derived from Eq. 2). The dots represent actual measurements. The average of $\cos 2\varphi$ (apolar order parameter) is calculated from the angular distribution function and the dimensionless guiding field is obtained by multiplying e.g. the applied electric field with the galvanotaxis coefficient.

The essential signal in the automatic controller of the bipolar cellular mode is the quadrupole moment of asymmetric spatial distribution of the loaded membrane-bound receptors. The bipolar mode of an amoeboid cell act as an automatic controllers where the symmetry axis of the bipolar mode is altered in small steps by means of the quadrupole moment of the asymmetric distribution of the loaded membrane-bound receptors.
Next, interacting and migrating cells are investigated.

Living Liquid Crystals

The single cell behavior was investigated as described above at low cell density. Now interacting cells without an applied extracellular guiding signal will be considered. A liquid crystal is expected if one arbitrary cell is guided by the submitted signals of the other cells. Two types of living liquid crystals are observed: A polar nematic liquid crystal [15] is obtained when the interacting cells produce a polar guiding field and an apolar nematic liquid crystal [16] for an apolar guiding field.

Polar Nematic Liquid Crystal

A condensed state formed by cells needs attractive cell-cell interaction. The vesicle fusion process is the essential feedback element in the intracellular signal chain but this process can also be used for the extracellular signaling if the vesicles contain a chemoattractant. Staved slime mold cells release c-AMP which attracts slime mold cells. Granulocytes attract each other if they are kept at low calcium concentration [15], but the released signal molecule of granulocytes at low calcium is still not yet known.
The formation of a liquid crystal phase is based on intrinsic properties of the migrating cells (automatic controller with an intrinsic noise source) and on the extracellular guiding signal which is produced by the other cells. It is difficult to sum up all the pair interactions. Therefore, the interaction of *one cell* with all the *others cells* can be approximated by introducing a mean guiding signal. Let us consider a cluster of dense backed but migrating cells. The cluster can grow since it attracts cells from its vicinity. The attractive signal should be proportional to the number of cells in one cluster. The cluster can shrink since cells from the boundary of the cluster diffuse away. A cluster of migrating cells is formed where the migrating cells point towards the center of the cluster (Fig. 3).

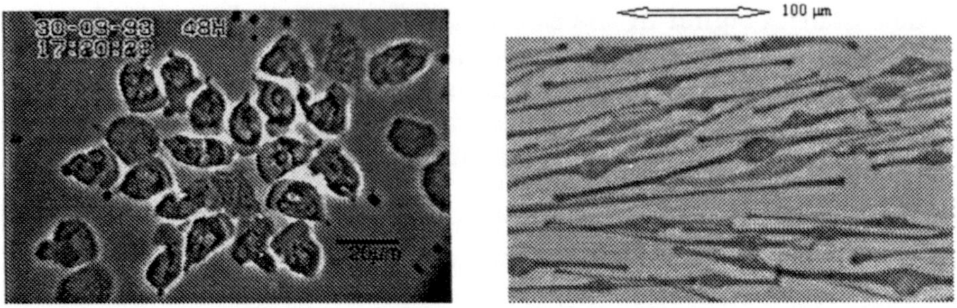

Figure 3: *Left side: granulocytes in the polar mode attract each other and form a cluster. The cells form a polar nematic liquid crystal. Right side: melanocyte in the bipolar mode orient their long axis in such a way that the dendrite-dendrite interaction is minimum. The cells form an apolar nematic liquid crystal.*

The cluster formation was a function of the mean cell density, ρ_0. A threshold behavior was predicted and actually observed. At low cell density ($\rho_0 < \rho_{th} = 150 - 300$ cells/mm^2) the migrating cells did not form clusters. Mainly single migrating cells were observed, but occasionally two or three cells stayed together for several minutes. At high cell density ($\rho_0 > \rho_{th}$), the migrating cells formed clusters with a very long lifetime ($1 - 2h$), but the clusters are not stable; they dissolve themselves for large times $t > 1 - 2h$.
The migrating and interacting cells in a cellular polar mode form a polar nematic liquid crystal. The condensed phase is a liquid where the center of mass of the cell show no order, however, the condensed phase with its elongated cells show a polar order.

Apolar Nematic Liquid Crystal

Amoeboid cells in the bipolar mode like to orient their long axis perpendicular to the extracellular guiding signal. One could say that the cells like to minimize the interaction with extracellular signals. With this knowledge in mind, one expects an apolar nematic liquid crystal for attracting cells having an elongated shape since the free volume is reduced if the cells orient their long axis in a preferred direction.
The machine equation for the orientation angle of a single cell is a stochastic equation as previously discussed (Eq. 2). The

extracellular guiding signal can now be exchanged by a mean field which is produced by all the other cells. One expects that the mean guiding field is proportional to the cell density, ρ, to the apolar order parameter, $\langle cos2\Theta\rangle$ and to $\sin 2\Theta$. The mean guiding field registrated by the cellular signal transformer, $g_2(\Theta, E_2)$, is then

$$g_2(\Theta, E_2) = \alpha\, \rho\, \langle cos2\Theta\rangle\, \sin 2\Theta \qquad (9)$$

with the coefficient α. The apolar order parameter can be predicted by introducing this function in eq. 4. This mean-field approach is actually verified for melanocytes [16]. For low cell density ($\rho < \rho_{th} \approx 110$ cells/mm^2) the stochastic process is stronger than the orienting process and the cells are randomly oriented. However, for high cell density ($\rho > \rho_{th}$) the orienting process is stronger than the stochastic process and a nematic liquid crystalline phase is expected as actually observed (Fig. 3). The experiments were performed from days up to several weeks [16]. The apolar order parameter as a function of the cell density is quite well described by the mean-field theory where the threshold concentration is the only fitting parameter.

Outlook

Fluid self-organized machines represent a new and important field in the triangle between physics, chemistry and life sciences. It is important e.g. in a cell, to know all steps of the biochemical reaction chain. In addition, one has to find the processes which creates the machine (coherence in the signal chain). The knowledge of the biochemical events and of the details of the working machine could lead to new frontiers. For example, phenomenon in cell biology are usually discussed on a molecular basis. The new concept is to discuss such phenomenon additionally on the basis of self-organized machines. It means in the case of pharmacology that compound design should not neglect the machine aspects. The self-organization of fluid systems to ordered structures plays an important role in morphogenesis, organogenesis, wound-healing, etc. Here, it is important to realize that the self-organized cellular system is much more sensitive to extracellular signals as the single cell. This knowledge opens new therapeutic possibilities.

Acknowledgement

I am grateful to Dr. Anne de Boisfleury-Chevance for her support and encouragement. This work was supported in part by "Fond der chemischen Industrie", "Deutsche Forschungsgemeinschaft", "Deutscher Akademischer Austauschdienst" and "Fondation de France".

References

[1] J.P. Trinkaus, *Cells into Organs* (Prentice-Hall Inc., Englewood Cliffs, 1984)

[2] P.C. Wilkinson, *Chemotaxis and Inflammation* (Churchill, London, 1974)

[3] H. Haken, *Synergetics* (Springer, Berlin, 1983)

[4] M.Schienbein, H. Gruler, *Phys. Rev. E*, **52**, 4183 (1995)

[5] H. Gruler, *Liq. Cryst.*, **24**, 49 (1997)

[6] S. E. Malawista, A. de Boisfleury-Chevance, *J. Cell Biol.*, **95**, 960 (1982)

[7] M.B. Hallet, *The Neutrophile: Cellular Biochemistry and Physiology*, CRC Press, Boca Raton, Florida

[8] M. Bessis, A. de Boisfleury-Chevance, *Nouv. Rev. Fr. Hematol.*, **11**, 377 (1971)

[9] H. Gruler, *Nouv. Rev. Fr. Hematol.*, **37**, 255 (1995)

[10] H. Gruler, *Chaos and Complexity* edited by J.T. Thanh Van (Editions Frontière, Gif sur Yvette) (1992)

[11] K. Franke and H. Gruler H, Eur. Biophys. J. **18**, 335 (1990)

[12] M. Schienbein, H. Gruler, *Phys. Rev. E*, **49**, 5462 (1994)

[13] H. Risken, *Fokker-Planck Equation* (Springer Verlag, Heidelberg, 1984)

[14] H. Gruler, in *Biologically Inspired Physics* edited by L. Peliti (Plenum Press, 1991)

[15] H. Gruler A. de Boisfleury-Chevance, *J.Phys. I France*, **4**, 1085 (1994)

[16] H. Gruler, U. Dewald, M. Eberhardt, to be published

The Metric Content of Spatial Views as Represented in the Primate Hippocampus

Alessandro Treves[1], Pierre Georges-Francois[2], Stefano Panzeri[2], Robert G Robertson[2] and Edmund T Rolls[2]

[1]SISSA - Programme in Neuroscience, 34014 Trieste, Italy
[2]Dept. of Exp. Psychology, Univ. of Oxford, Oxford OX1 3UD, UK

Abstract. Coexisting memory representations of the same information may differ in the amount of structure they embody, i.e. in the metric of relationships among individual memory items. Such an amount of structure may be quantified by the metric content index. We extract the metric content of the representation of spatial views in the monkey hippocampus and parahippocampal cortical areas, and find indications of quantitative differences that might be associated with the connectivity pattern in different neural substrates.

Keywords. spatial views, hippocampus, information theory, decoding, ultrametricity, metric structure

1 Quantifying the amount of structure

The structure of neural representations of the outside world has been studied in detail in some simple situations. Typically these are situations in which a well defined correlate of neuronal activity (i.e. a stimulus, a response, or even a behavioural state) is characterized by one or a few parameters that are made to vary continuously or in steps. Examples are the Hubel and Wiesel [2] description of orientation selectivity in cat visual cortex, the O'Keefe [5] finding of place cells in the rat hippocampus, the mitral cell coding of n-aliphatic acid hydrocarbon length in the olfactory system [12], the coding of the direction of movement in 3D-space in the primate motor cortex [1].

In many interesting situations, though, especially in those parts of the brain which are more remote from the periphery, external correlates, or, as we shall refer to them for simplicity, stimuli, do not vary (either continuously, or in steps) along any obvious physical dimension. Often, in experiments, the set of stimuli used is just a small ensemble of a few disparate individual items, arbitrarily selected and difficult to classify systematically. Examples for the ventral visual system are faces [9], simple or complex [4] abstract patterns, or the schematic objects reached with the reduction procedure of Tanaka *et al* [13]. In such situations, the resulting patterns of neuronal activity across populations of cells can still provide useful insight on the structure of neuronal representations of the outside world, but such insight has to be derived independently of any explicit correlation with a natural, physical structure of the stimulus set.

The only obvious *a priori* metric of the stimulus set, in the general case, is the trivial categorical metric of each element s being equal to itself, and different from any other element in the set. *A posteriori*, the neuronal firing patterns embed the stimulus set into a potentially metric structure defined by the similarities and differences among the patterns corresponding to the various elements. A truly metric structure can be extracted by quantifying such similarities and differences into a notion of distance (among firing patterns) that satisfies the 3 required relations: positivity, symmetry, the triangle inequality. At a more basic level, though, the overall amount of structure, i.e. the overall importance of relations of similarity and difference among firing patterns, can be quantified even independently of any notion of distance, just from a matrix $Q(s|s')$ characterizing the confusability of s' with s, a matrix which need not be symmetrical. It is moreover important to notice that such a matrix $Q(s|s')$ can indeed be derived, as discussed below, from the firing patterns corresponding to each stimulus s, but it can also be derived from other, e.g. behavioural, measures. Behavioural measures of the confusability of s' with s do not access the representation of the two stimuli directly, but indirectly they reflect the multiplicity of neural representations that are important in generating that particular behaviour. If some of these representations are damaged or lost, as in brain-damaged patients, the resulting behavioural measures can be indicative of the structure of the surviving representations [3].

The amount of structure can be quantified by comparing the mutual information in the matrix $Q(s|s')$,

$$I = \sum_{s,s' \in S} Q(s|s')P(s') \log_2 \frac{Q(s|s')}{\sum_{s''} Q(s|s'')P(s'')} \tag{1}$$

with its minimum and maximum values I_{min} and I_{max} [14] corresponding to a given percent correct $f_{cor} = \sum_s Q(s|s)P(s)$. The lowest information values compatible with a given f_{cor} are those attained when equal probabilities (or equal frequencies of confusion) result for all wrong stimuli. In this case one finds

$$I_{min} = \log_2 S + f_{cor} \log_2 f_{cor} + (1 - f_{cor}) \log_2[(1 - f_{cor})/(S - 1)]. \tag{2}$$

Conversely, maximum information for a given f_{cor} is contained in the confusion matrix when stimuli are confused only within classes of size $1/f_{cor}$ (for analytical simplicity we assume that each class may contain a non integer number of elements), and the individual stimuli within the class are allocated on a purely random basis. It is easy to see that then

$$I_{max} = \log_2 S + \log_2 f_{cor}. \tag{3}$$

Interpreting the probability of confusion as a monotonically decreasing function of some underlying distance (e.g. as discussed above), the first situation can be taken to correspond to the limit in which the stimuli form an equilateral simplex, or equivalently the stimulus set is drawn from a space of extremely high dimensionality. In the Euclidean $d \to \infty$ limit, points drawn at random from a finite e.g. hyperspherical region tend to be all at the same distance from each other, and from the point of view of the metric of the set this is the *trivial* limit mentioned above. The second situation can be taken to correspond to the *ultrametric* limit, instead, in which all stimuli at distance less than a critical value from each other form clusters such that all distances between members of different classes are above the critical value. This is a non-Euclidean structure (although it could be embedded in a Euclidean space of sufficiently large dimension), and it is a first example of the possible emergence of non-Euclidean aspects from a quantitative analysis that does not rely on *a priori* assumptions.

Intermediate situations between the two extremes are easy to imagine and can be parametrized in a number of different ways. A convenient parameter that simply quantifies the relative amount of information in excess of the minimum, without having to assume any specific parametrization for the confusion matrix, is

$$\lambda = \frac{I - I_{min}}{I_{max} - I_{min}} \tag{4}$$

which ranges from 0 to 1 and can be interpreted as measuring the *metric content* of the matrix. What is quantified by λ can be called the metric content not in the sense that it requires the introduction of a real metric, but simply because it gives the degree to which relationships of being close or different (distant), among stimuli, emerge in the $Q(s|s')$ matrix. For $\lambda = 0$ such relationships are irrelevant, to the point that if confusion occurs, it can be with any (wrong) stimulus. For $\lambda = 1$ close stimuli are so similar as to be fully confused with the correct one, whereas other stimuli are 'maximally distant' and never mistaken for it.

In summary, the metric content index λ quantifies the dispersion in the distribution of errors, from maximal, $\lambda = 0$, to minimal, $\lambda = 1$. The errors may be actual behavioural errors in identifying or categorizing stimuli or in producing appropriate responses, or errors which a neuronal population appears to be making, as an outside observer infers by reading out the spiking activity of the population. We now turn to what is exactly implied by the notion of 'reading out'.

2 Decoding the responses of spatial view cells

Decoding the spike trains emitted by a population of neurons, when a stimulus s from a given set is presented, means applying an algorithm that estimates, given the current spike trains $\vec{r}_s$ and those previously recorded in response to each stimulus, the likelihoods for each (s') of the possible stimuli to be the current one, $L(s'|\vec{r}_s)$. The stimulus $s' = s_p$ for which this likelihood is maximal can be said to be the stimulus *predicted* on the basis of the response. In general s_p will not coincide with the true s and the accuracy in the decoding can of course be measured by the percent correct decoding (or the corresponding fraction f_{cor}), but also by the mutual information in the joint probability table $Q(s, s_p)$,

$$I = \sum_{s,s_p \in S} Q(s, s_p) \log_2 \frac{Q(s, s_p)}{P(s)Q(s_p)}. \tag{5}$$

This is the quantity referred to above. It measures the information in the predictions based on maximum likelihood, and as such it does not only reflect, like percent correct, the number of times the decoding is exact, but also, beyond percent correct, the distribution of wrong decodings. A further quantity which it is sometimes useful to consider is the mutual information

$$I_p = \sum_{s,s' \in S} P(s, s') \log_2 \frac{P(s, s')}{P(s)P(s')} \tag{6}$$

obtained from the *probability* $P(s'|s)$ of confusing s with s', which is given by averaging $L(s'|\vec{r}_s)$ over the responses to s. This second information measure reflects, unlike the first, also the degree of certainty with which each single trial has been decoded, and it thus sheds light on a further aspect of the quality attained in decoding. Both information quantities suffer from limited sampling distortions [15, 6] but the second much less than the first, in the sense that, with the limited sampling correction procedures we have developed, I_p can be estimated accurately even with few trials per stimulus, while I requires more trials. Hovever in practice, especially when extracting these measures from limited periods of firing of cortical cells, I is a much better estimate of the actual information contained in the firing (i.e., before decoding) than I_p [7], and because of this fidelity it is preferable to rely on measures of I whenever limited sampling distortions are not the main concern. We note that the metric content index appropriate to I_p would be derived in the same terms, by only replacing f_{cor} with the analogous quantity based on probabilities, $g_{cor} = \sum_s P(s|s)P(s)$.

Decoding algorithms can be optimised to extract as much information as possible, or they can be modelled on the decoding likely to be implemented by real neurons downstream of the recorded populations. Information and percent correct values in the decoding of face cells responses from the primate temporal visual cortex have been reported [10]. There we show that simple, neuronally plausible decoding algorithms, based on dot product operations, perform virtually like optimal decoding algorithms in terms of I, and are only 20-30% inferior in terms of I_p. This is because the simple dot product algorithms are poorer at quantifying likelihoods, even if they order them correctly and identify correctly the most likely stimulus that can be predicted for each trial.

f_{cor}, I and I_p all depend on the number of cells in the population, as recording the responses of more cells obviously allows better decoding. We have reported the important result [10] that the information decoded from face cells appears to grow linearly with the number of cells in the population, until it begins to saturate at the maximum allowed, which is just the entropy of the stimulus set, $H = -\sum_s P(s) \log_2 P(s)$. This result implies that the different cells in the sample tend to code for different aspects of the stimulus set, so that each contributes an additive term to the information provided by the population. This result appears to hold for the data recorded in a number of experiments, including both the primate inferior temporal cortex face cells [10] mentioned above and the primate hippocampal spatial view cells [11] considered in this report, but also primate orbitofrontal cells coding for odours (Rolls, Treves and Critchley, in preparation), rat hippocampal cells coding for spatial position (e.g. [16]; and also [17]), and rat somatosensory cells coding for whisker deflection [8].

The issue we want to focus on here is not, however, how the accuracy in the decoding depends on the number of cells in the population, but rather how it provides insight on the structure of the stimulus set as encoded in the firing of different populations of cells, and as quantified by the λ metric content index.

3 The metric content in neighbouring areas

The data we consider are the responses of spatial view cells in the primate hippocampus, described by Rolls et al [11], to which we refer for all the details of the experiment and analysis. Briefly, single cells were successively recorded in 2 monkeys while the animals were free to locomote in the lab, and their gaze direction was simultaneously recorded with magnetic coils. For the purpose of the analysis, gaze directions were discretized into 16 'views', which corresponded to an equal number of portions of the lab walls. 20 of the cells used here were recorded in one anymal (monkey *av*) and 6 in another (monkey *az*). Pseudosimultaneous response vectors were constructed by randomly pairing equal numbers of trials in which each cell included in the vectors was recorded in its response to any given view.

Each trial consisted of a 100ms long stretch during which the monkey's gaze was fixed within one of the 16 preset spatial views. At the end of each trial, if the gaze remained fixed for another 100ms period, another trial associated with the same spatial view was constructed, and so on. Decoding was thus based on the number of spikes emitted by each cell considered in the current sample, within one of the 100ms pseudosimultaneous trials. Typically about 80 trials were available for each view.

Note the difference between our response vectors and Georgopoulos' pop-

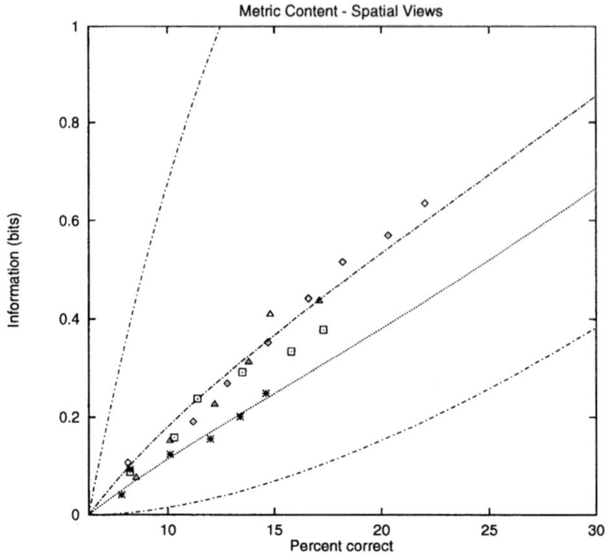

Figure 1: The information decoded from different cell populations *vs.* the corresponding percent correct. I_{min} and I_{max} are indicated with dash-dotted lines, along with the curve $\lambda = 0.25$. The other curve is $\lambda = 0.15$. Datapoints are for populations of CA3 ($*$), CA1 (triangles), parasubiculum (squares) and parahippocampal gyrus cells (diamonds).

ulation vectors [1], which live in the physical 3D or 2D movement space rather than in the space of dimensionality equal to the number of cells included, and which correspond to a continuous rather than a discretized correlate. Vectors were constructed including all 26 cells, only the cells recorded in each animal, only those in a given brain region of both animals, and so on exploring different combinations. Since the metric content index is expected to be relatively constant as the number of cells randomly sampled from an homogeneous population varies, comparisons can be made, to some extent, even among the metric content characterizing vectors of different dimensionality.

One can see from the figure the extent to which metric content, while not being a strictly invariant characterization of the responses, valid for all percent correct values, is still a relatively stable index. For each given cortical area, as more cells are considered, both percent correct and decoded information grow, and the relation between the two, expressed as metric content, varies somewhat in a limited band of values characterizing each cortical area. One should note that the variability as the number of cells varies is limited only because of the extensive averaging we perform, e.g. when decoding from 3 CA1 cells, over nearly all possible triplets of cells from the 6 available from the CA1 area. Obviously, this averaging cannot compensate from the fluctuations induced by the very limited number of cells – 6 – in the total CA1 sample. This is ultimately one of the main limits of this preliminary analysis, which prevents us from drawing definite conclusions.

The figure shows the individual datapoints obtained for the average sample of c cells from each cortical area, and also two representative lines of constant metric content, one for $\lambda = 0.25$ and one for $\lambda = 0.15$. Datapoints from hippocampal area CA1 ($c = 6$), from the parahippocampal gyrus (PHG, $c = 8$) and from the parasubiculum (PSUB, $c = 6$) tend to cluster around the upper metric content curve, while datapoints for hippocampal area CA3 ($c = 6$) tend to cluster around the lower curve. As it happened, 4 out of 6 CA3 cells were recorded in monkey *az*, while all but 2 (1 CA1 and 1 PSUB) of the other cells were from monkey *av*. Extensive testing with subsets of cells taken from both the same area *and* the same monkey failed to clarify conclusively whether the emerging metric content difference is due to the area or to the monkey.

4 Comments and outlook

The data analysed in this paper are not fully adequate, on at least two accounts. First, the number of cells recorded and the number of 100ms trials available for each cell and each spatial view were not sufficiently large to safely avoid limited sampling effects. Second, the recordings should be simultaneous, and from the same monkey, to avoid differences due to slow changes in the representations e.g. with learning or habituation or increasing boredom, or due to individual differences. Both inadequacies can be removed with parallel recording from several cells at once, so the preliminary results of the type of analysis presented here will soon be confirmed or disproved by analysing more adequate data.

Within these limits, one possible interpretation of the different metric content in the CA3 area, with respect to the other 3 areas sampled, lies in the different pattern of connectivity, whereby in CA3 recurrent collateral connections are the numerically dominant source of inputs to pyramidal cells, and travel relatively long distance, to form an extended network connected by intrinsic circuitry. Considerations based on simplified network models suggest that such a connectivity pattern would express memory representations with a different metric structure from those expressed by networks of different types. The difference could be further related to the qualitative nature of the memory representation, which might be characterized as being more *episodic* in CA3 and more *structured* in the other areas. The metric content depends also on the average sparseness of these representations, though, and further analyses are required to dissociate the effects of connectivity (and of representational structure) from those purely due to changes in sparseness.

The present recordings were from neighbouring areas in the temporal lobes, and it is possible that any difference among memory representations will be more striking when more distant areas are compared. In addition, it is possible that any difference may be more striking when the correlate considered does not have its own intrinsic metric, as with spatial views, but instead lives in a high dimensional space, as e.g. with faces, thereby letting more room for arbitrary metric structures to be induced in the neural representations by the learning process. For both reasons, it will be interesting to extend this analysis to entirely different experiments, sharing with the present one only the generic requirement that different populations of cells are recorded in their response to the same set of stimuli, or in general correlates.

Finally, possible changes in the representations that develop with time could be examined by recording from the *same* populations – not the same cells – over periods during which some behaviourally relevant phenomenon may have occured, such as new learning, forgetting, or a modulation of the existing representations. One specific such modulation of interest for the case of human patients is the one resulting from localized lesions to another cortical area, which may affect the structure of the representations in surviving areas of the cortex.

Acknowledgements

Partial support was from the Medical Research Council of the UK, the Human Capital and Mobility Program of the EU, and the National Research Council of Italy.

References

[1] Georgopoulos, A.P., Kettner, R.E. and Schwartz, A.B.: Primate motor cortex and free arm movements to visual targets in three-dimensional space. II. Coding of the direction of movement by a neuronal population. J. Neurosci. *8*, 2928-2937 (1988)

[2] Hubel, D.H. and Wiesel, T.N.: Sequence regularity and geometry of orientation columns in the monkey striate cortex. J. Comp. Neurol. **158**, 267-294 (1974)

[3] Lauro-Grotto, R., Piccini, C., Borgo, F. and Treves, A.: What remains of memories lost in Alzheimer and herpetic encephalitis. Soc. Neurosci. Abs. **23**, 734.2 (1997)

[4] Miyashita, Y. and Chang, H.S.: Neuronal correlate of pictorial short-term memory in the primate temporal cortex. Nature **331**, 68-70 (1988)

[5] O'Keefe, J.: A review of the hippocampal place cells. Progr. Neurobiol. **13**, 419-439 (1979)

[6] Panzeri, S. and Treves, A.: Analytical estimates of limited sampling biases in different information measures. Network **7** 87-107 (1996)

[7] Panzeri, S., Treves, A., Schultz, S. and Rolls, E.T.: On decoding the responses of a populations of neurons from short time windows. Oxford U., Dep. of Exp. Psych. preprint (1998)

[8] Petersen, R.S., Treves, A., Lebedev, M. and Diamond, M.: Information theoretic analysis of the responses of rat cortical neurons to vibrissal stimulation. Soc. Neurosci. Abs **23**, 913.17 (1997)

[9] Rolls, E.T.: Neurophysiological mechanisms underlying face processing within and beyond the temporal cortical visual areas. Phil. Trans. Roy. Soc. **B 335**, 11-21 (1992)

[10] Rolls, E.T., Treves, A. and Tovee, M.J.: The representational capacity of the distributed encoding of information provided by populations of neurons in the primate temporal visual cortex. Exp. Brain Res. **114**, 149-162 (1997)

[11] Rolls, E.T., Treves, A., Robertson, R.G., Georges-Francois, P. and Panzeri, S.: Information about spatial view in an ensemble of primate hippocampal cells. J. Neurophysiol. **79**, 1797-1813 (1998)

[12] Sullivan, S.L. and Dryer, L.: Information processing in mammalian olfactory system. J. Neurobiol. **30**, 20-36 (1996)

[13] Tanaka, K.: Neuronal mechanisms of object recognition. Science **262**, 685-688 (1993)

[14] Treves, A.: On the perceptual structure of face space. Biosystems **40**, 189-196 (1997)

[15] Treves, A. and Panzeri, S.: The upward bias in measures of information derived from limited data samples. Neural Comp. **7**, 399-407 (1995)

[16] Treves, A., Skaggs, W.E. and Barnes, C.A.: How much of the hippocampus can be explained by functional constraints? Hippocampus **6**, 666-674 (1996)

[17] Wilson, M. and McNaughton, B.L.: Dynamics of the hippocampal ensemble code for space. Science **261**, 1055-1058 (1993)

NATO ASI Series F

Including Special Programmes on Sensory Systems for Robotic Control (ROB) and on Advanced Educational Technology (AET)

Vol. 145: Advanced Educational Technology: Research Issues and Future Potential. Edited by T. T. Liao. VIII, 219 pages. 1996. *(AET)*

Vol. 146: Computers and Exploratory Learning. Edited by A. A. diSessa, C. Hoyles and R. Noss. VIII, 482 pages. 1995. *(AET)*

Vol. 147: Speech Recognition and Coding. New Advances and Trends. Edited by A. J. Rubio Ayuso and J. M. López Soler. XI, 505 pages. 1995.

Vol. 148: Knowledge Acquisition, Organization, and Use in Biology. Edited by K. M. Fisher and M. R. Kibby. X, 246 pages. 1996. *(AET)*

Vol. 149: Emergent Computing Methods in Engineering Design. Applications of Genetic Algorithms and Neural Networks. Edited by D.E. Grierson and P. Hajela. VIII, 350 pages. 1996.

Vol. 150: Speechreading by Humans and Machines. Edited by D. G. Stork and M. E. Hennecke. XV, 686 pages. 1996.

Vol. 151: Computational and Conversational Discourse. Burning Issues – An Interdisciplinary Account. Edited by E. H. Hovy and D. R. Scott. XII, 202 pages. 1996.

Vol. 152: Deductive Program Design. Edited by M. Broy. IX, 467 pages. 1996.

Vol. 153: Identification, Adaptation, Learning. Edited by S. Bittanti and G. Picci. XIV, 553 pages. 1996.

Vol. 154: Reliability and Maintenance of Complex Systems. Edited by S. Özekici. XI, 589 pages. 1996.

Vol. 155: Cooperation: Game-Theoretic Approaches. Edited by S. Hart and A. Mas-Colell. VIII, 328 pages. 1997.

Vol. 156: Microcomputer-Based Labs: Educational Research and Standards. Edited by R.F. Tinker. XIV, 398 pages. 1996. *(AET)*

Vol. 157: Logic of Computation. Edited by H. Schwichtenberg. VII, 396 pages. 1997.

Vol. 158: Mathematical Methods in Program Development. Edited by M. Broy and B. Schieder. VIII, 528 pages. 1997.

Vol. 159: Fractal Image Encoding and Analysis. Edited by Y. Fisher. XIV, 368 pages. 1998.

Vol. 160: Discourse, Tools, and Reasoning: Essays on Situated Cognition. Edited by L.B. Resnick, R. Säljö, C. Pontecorvo and B. Bunge. XII, 474 pages. 1997. *(AET)*

Vol. 161: Computational Methods in Mechanical Systems. Edited by J. Angeles and E. Zakhariev. X, 425 pages. 1998.

Vol 162: Computational Intelligence: Soft Computing and Fuzzy-Neuro Integration with Applications. Edited by O. Kaynak, L.A. Zadeh, B. Türkşen, I.J. Rudas. IX, 538 pages. 1998.

Vol. 163: Face Recognition: From Theory to Applications. Edited by H. Wechsler, P.J. Phillips, V. Bruce, F. Fogelman Soulié, T.S. Huang. IX, 626 pages. 1998.

Vol. 164: Workflow Management Systems and Interoperability. Edited by A. Doğaç, L. Kalinichenko, M.T. Özsu, A. Sheth. XVII, 481 pages. 1998.

Vol. 166: Operations Research and Decision Aid Methodologies in Traffic and Transportation Management. Edited by M. Labbé, G. Laporte, K. Tanczos and P. Toint. XIII, 339 pages. 1998.

Vol. 167: Neural Circuits and Networks. Edited by V. Torre and J. Nicholls. VIII, 247 pages, 1998.

Vol. 168: Generalization in Neural Networks and Machine Learning. Edited by Ch.M. Bishop. XII, 353 pages, 1999.

MIX
Papier aus verantwortungsvollen Quellen
Paper from responsible sources
FSC® C105338
FSC
www.fsc.org

If you have any concerns about our products,
you can contact us on
ProductSafety@springernature.com

In case Publisher is established outside the EU,
the EU authorized representative is:
Springer Nature Customer Service Center GmbH
Europaplatz 3, 69115 Heidelberg, Germany

Printed by Libri Plureos GmbH
in Hamburg, Germany